Optical Properties ar
Applications of Semiconductors

Optical Properties and Applications of Semiconductors

Edited by

Inamuddin
Mohd Imran Ahamed
Rajender Boddula
Tariq Altalhi

CRC Press
Taylor & Francis Group
Boca Raton London New York

CRC Press is an imprint of the
Taylor & Francis Group, an **informa** business

First edition published 2023
by CRC Press
6000 Broken Sound Parkway NW, Suite 300, Boca Raton, FL 33487-2742

and by CRC Press

4 Park Square, Milton Park, Abingdon, Oxon, OX14 4RN

CRC Press is an imprint of Taylor & Francis Group, LLC

Library of Congress Cataloging-in-Publication Data

Names: Inamuddin, 1980- editor. | Ahamed, Mohd Imran, editor. | Boddula,
  Rajender, editor. | Altalhi, Tariq, editor.
Title: Optical properties and applications of semiconductors / edited by
  Inamuddin, Mohd Imran Ahmed, Rajender Boddula, Tariq A. Altalhi.
Description: First edition. | Boca Raton, FL : CRC Press, 2022. | Includes
  bibliographical references and index.
Identifiers: LCCN 2021060504 (print) | LCCN 2021060505 (ebook) | ISBN
  9781032036984 (hbk) | ISBN 9781032037028 (pbk) | ISBN 9781003188582
  (ebk)
Subjects: LCSH: Semiconductors--Optical properties.
Classification: LCC QC611.6.O6 O6627 2022 (print) | LCC QC611.6.O6
  (ebook) | DDC 537.6/226--dc23/eng20220412
LC record available at https://lccn.loc.gov/2021060504
LC ebook record available at https://lccn.loc.gov/2021060505

ISBN: 978-1-032-03698-4 (hbk)
ISBN: 978-1-032-03702-8 (pbk)
ISBN: 978-1-003-18858-2 (ebk)

DOI: 10.1201/9781003188582

Typeset in Times
by KnowledgeWorks Global Ltd.

Contents

Preface

Semiconductors with optical characteristics have found widespread use in evolving semiconductor photovoltaics, where optical features are important. The industrialization of semiconductors and their allied applications have paved the way for optical measurement techniques to be used in new ways. Semiconductors have demonstrated a wide range of optical properties, electrical band structure, and energy levels. Due to their unique properties, semiconductors are key components in daily employed technologies in health care, computing, communications, green energy, and in a range of other uses. In addition, solar fuels, photovoltaics, sensors, wireless-remote controllers, laser diodes, optical communications, photodetectors, optoelectronics, phototransistors, photodiodes, and light-emission devices are some of the few potential uses of semiconductors. Hence, the practical application of semiconductors demands the understanding of their optical properties by the varied industrial and research groups.

Optical Properties and Applications of Semiconductors is intended to elaborate classification, fundamental optical properties, and applications of semiconductors. It summarizes the information, as well as the optical characteristics and applicability of semiconductors, through an in-depth review of the literature. It brings together people from the domain of industry specialists as far as semiconductor unique properties and applications are concerned. It is an invaluable guide to engineers, students, professors, scientists, and R&D industrial experts working in the field of applied physics concerning semiconductors' opto-characteristics and applications. The summaries of the work reported in the following 10 chapters are as follows:

Chapter 1 addresses the fabrication routes of semiconductor optical fibers. Two methods of core melting and modified chemical vapor deposition are discussed in detail. In addition, the structural and optical properties of semiconductor optical fibers are summarized. Finally, the application and prospects of semiconductor optical fibers are discussed.

Chapter 2 focuses on intelligent semiconducting materials for solar photocatalysis and their optical properties. These materials provide the tuned bandgap, surface functionalization, and desired range of irradiation by tuning the optical properties. It also discusses the hybrid semiconducting materials that lie in the visible region for effective solar photocatalysis with their fabrication methods.

Chapter 3 outlines the potential role of photo-responsive semiconductors for optical memory device applications. It also discusses the optoelectronic properties using theoretical techniques. The basic aim is to highlight the benefits and future viability of photo-responsive semiconductors for optical memory device applications that have been documented in the literature.

Chapter 4 highlights the different classes of semiconductor photocatalysts. Various semiconductors widely used for agricultural purposes are discussed in detail. It provides an overview of the semiconductors and their implemented properties to enhance agricultural productivity.

Chapter 5 elaborates the nonlinear optics theory as well as nonlinear effects like resonant/nonresonant, photothermal and photoelectric nonlinearities, nonlinear refraction and absorption. Harmonic generation up to third order in semiconductors is deliberated through Maxwell equations along with relation of susceptibility. Applications of nonlinear organic semiconductors are also briefly discussed.

Chapter 6 is a discussion about semiconductor photoresistors, their types, and characteristics. The applications of photoresistors in image sensing, as light sensors, and in wearable electronic gadgets are briefly discussed. New materials for photoresistor production, as well as the development of their uses in other industries, must be identified.

Chapter 7 showcases the development of present photovoltaic technology. It emphasizes different photovoltaic semiconductor materials utilized for solar energy applications such as crystalline, amorphous, thin-film, and next-generation photovoltaics. The main aim here is to discuss the efficiency improvement, stability, and cost-effectiveness of solar module production using these semiconductor materials.

Chapter 8 emphasizes various optical properties of semiconductors for the development of solar photocatalysis. The principle of semiconductor photocatalysis, influencing factors, and types of photocatalysts are discussed in detail. The major focus is to outline the available methods for improving the efficiency of developing efficient photocatalysts for smart and sustainable energy production.

Chapter 9 deliberates the basic theory of linear optics as well as different linear optical properties of semiconductors by using Maxwell equations in precise detail, by considering the behavior of light in medium/vacuum. Effects of external magnetic cum electric fields are also studied in bulk along with low dimensional structures.

Chapter 10 describes the optical behaviors of different metal oxide semiconductors using theoretical tools. The first-principle calculations with various modifications and approaches are presented to enhance the computing optical properties of metal oxide semiconductors.

Editors

Inamuddin, PhD, is an Assistant Professor in the Department of Applied Chemistry, Aligarh Muslim University, Aligarh, India. He earned his M.Sc. in organic chemistry at Chaudhary Charan Singh (CCS) University, Meerut, India, in 2002. He earned his M.Phil. and PhD in Applied Chemistry at Aligarh Muslim University (AMU), India, in 2004 and 2007, respectively. He has extensive research experience in the multidisciplinary fields of analytical chemistry, materials chemistry, electrochemistry, and, more specifically, renewable energy and environment. He has worked on different research projects as project fellow and senior research fellow funded by the University Grants Commission (UGC), Government of India, and the Council of Scientific and Industrial Research (CSIR), Government of India. He has received the Fast Track Young Scientist Award from the Department of Science and Technology, India, to work in the area of bending actuators and artificial muscles. He has completed four major research projects sanctioned by the University Grant Commission, Department of Science and Technology, the Council of Scientific and Industrial Research, and the Council of Science and Technology, India. He has published 196 research articles in international journals of repute and 19 book chapters in knowledge-based book editions published by renowned international publishers. He has published 150 edited books with Springer (UK), Elsevier, Nova Science Publishers, Inc. (USA), CRC Press – Taylor & Francis Asia Pacific, Trans Tech Publications Ltd. (Switzerland), IntechOpen Limited (UK), Wiley-Scrivener (USA) and Materials Research Forum LLC (USA). He is a member of various journals' editorial boards. He is an Associate Editor for several journals (*Environmental Chemistry Letter, Applied Water Science* and *Euro-Mediterranean Journal for Environmental Integration*, Springer-Nature), Frontiers Section Editor (*Current Analytical Chemistry*, Bentham Science Publishers), Editorial Board Member (*Scientific Reports*, Nature), Editor (*Eurasian Journal of Analytical Chemistry*), and Review Editor (*Frontiers in Chemistry*, Frontiers, UK). He has also guest-edited various thematic special issues to the journals of Elsevier, Bentham Science Publishers, and John Wiley & Sons, Inc. He has attended as well as chaired sessions at various international and national conferences. He has worked as a Postdoctoral Fellow, leading a research team at the Creative Research Initiative Center for Bio-Artificial Muscle, Hanyang University, South Korea, in the field of renewable energy, especially biofuel cells. He has also worked as a Postdoctoral Fellow at the Center of Research Excellence in Renewable Energy, King Fahd University of Petroleum and Minerals, Saudi Arabia, in the field of polymer electrolyte membrane fuel cells and computational fluid dynamics of polymer electrolyte membrane fuel cells. He is a life member of the *Journal of the Indian Chemical Society*. His research interests include ion exchange materials, a sensor for heavy metal ions, biofuel cells, supercapacitors and bending actuators.

Mohd Imran Ahamed, PhD, is working as Research Associate at the Department of Chemistry, Aligarh Muslim University (AMU), Aligarh, India. He received his B.Sc. (Hons) Chemistry and Ph.D. (Chemistry) degrees from AMU. He has completed his M.Sc. (Organic Chemistry) from Dr. Bhimrao Ambedkar University, Agra, India. He has published several research and review articles in various international scientific journals. He has co-edited 57 books with Springer (UK), Elsevier, CRC Press – Taylor & Francis Asia Pacific, Materials Research Forum LLC (USA) and Wiley-Scrivener (USA). His research work includes ion-exchange chromatography, wastewater treatment, and analysis, bending actuator and electrospinning.

Rajender Boddula, PhD, works with the Chinese Academy of Sciences – President's International Fellowship Initiative (CAS-PIFI) at the National Center for Nanoscience and Technology (NCNST, Beijing). He earned an MS in organic chemistry at Kakatiya University, Warangal, India, in 2008.

He earned his PhD in chemistry with highest honors in 2014 for the work titled "Synthesis and Characterization of Polyanilines for Supercapacitor and Catalytic Applications" at the CSIR-Indian Institute of Chemical Technology (CSIR-IICT) and Kakatiya University (India). Before joining the National Center for Nanoscience and Technology (NCNST) as CAS-PIFI research fellow, China, Dr Boddula worked as a senior research associate and postdoc at National Tsing-Hua University (NTHU, Taiwan) respectively in the fields of bio-fuel and CO_2 reduction applications. His academic honors include a University Grants Commission National Fellowship and many merit scholarships, study-abroad fellowships from Australian Endeavour Research Fellowship, and CAS-PIFI. He has published many scientific articles in peer-reviewed international journals, authored around 20 book chapters, and also served as an editorial board member and a referee for reputed international peer-reviewed journals. He has published edited books with Springer (UK), Elsevier, Materials Science Forum LLC (USA), Wiley-Scrivener (USA), and CRC Press – Taylor & Francis Group. His specialized areas of research are energy conversion and storage, which include sustainable nanomaterials, graphene, polymer composites, heterogeneous catalysis for organic transformations, environmental remediation technologies, photoelectrochemical water-splitting devices, biofuel cells, batteries and supercapacitors.

Tariq Altalhi, PhD, joined the Department of Chemistry at Taif University, Saudi Arabia, as Assistant Professor in 2014. He received his doctorate degree from University of Adelaide, Australia, in the year 2014 with Dean's Commendation for Doctoral Thesis Excellence. He was promoted to the position of the Head of Chemistry Department at Taif University in 2017 and Vice Dean of Science College in 2019 till now. In 2015, one of his works was nominated for GreenTec Awards (based in Germany), Europe's largest environmental and business prize, amongst top ten entries. He has co-edited various scientific books. His group is involved in fundamental multidisciplinary research in nanomaterial synthesis and engineering, characterization, and their application in molecular separation, desalination, membrane systems, drug delivery, and biosensing. In addition, he has established key contacts with major industries in Kingdom of Saudi Arabia.

Contributors

Aamir Ahmed
Department of Physics
University of Jammu
Jammu, Jammu and Kashmir, India

Jazib Ali
Electronic Engineering Department
University of Rome
Tor Vergata, Italy

Syed Wazed Ali
Department of Textile and Fibre
    Engineering
Indian Institute of Technology
New Delhi, India

Hafeez Anwar
Department of Physics
University of Agriculture
Faisalabad, Pakistan

Sandeep Arya
Department of Physics
University of Jammu
Jammu, Jammu and Kashmir, India

Ghulam Abbas Ashraf
Department of Physics
Zhejiang Normal University
Jinhua, China

Asma Ayub
Department of Physics
University of the Punjab
Lahore, Pakistan

Satyaranjan Bairagi
Department of Textile and Fibre Engineering
Indian Institute of Technology
New Delhi, India

Prasun Banerjee
Multiferroic and Magnetic Material Research
    Laboratory (MMMRL)
Gandhi Institute of Technology and
    Management (GITAM) University
Bengaluru, Karnataka, India

Swagata Banerjee
Department of Textile and Fibre Engineering
Indian Institute of Technology
New Delhi, India

Zhijun Du
Department of Design
Seoul National University
    Seoul, Republic of Korea

Seerat Fatima
Department of Physics
University of the Punjab
Lahore, Pakistan

Mridula Guin
Department of Chemistry and Biochemistry
Sharda University
Noida, India

Rizwan Haider
Department of Chemical Engineering
Shanghai Jiao Tong University
Shanghai, China

Fayyaz Hussain
Department of Physics
Materials Simulation Research
    Laboratory (MSRL)
Bahauddin Zakariya University
Multan, Pakistan

Iqra Ilyas
Department of Physics
University of the Punjab
Lahore, Pakistan

Muhammad Imran
Department of Physics
Government College University Faisalabad
Faisalabad, Pakistan

Sapana Jadoun
Department of Analytical and Inorganic
    Chemistry
Faculty of Chemical Sciences
University of Concepción
Concepción, Chile

Fahmeeda Kausar
Department of Chemical Engineering
Shanghai Jiao Tong University
Shanghai, China

Rana Muhammad Arif Khalil
Department of Physics
Materials Simulation Research
    Laboratory (MSRL)
Bahauddin Zakariya University
Multan, Pakistan

Kiran Kumar Kondamareddy
Department of Physics
School of Pure Sciences
College of Engineering, Science and
    Technology
Fiji National University, Natabua Campus
Lautoka, Fiji

Tanaya Kundu
Department of Chemistry
Indian Institute of Science
Bangalore, India

Dinesh Rangappa
Department of Applied Sciences
Centre for Post Graduate Studies
Visvesvaraya Technological University
Bengaluru, Karnataka, India

Umbreen Rasheed
Department of Physics
Materials Simulation Research
    Laboratory (MSRL)
Bahauddin Zakariya University
Multan, Pakistan

Dhirendra Singh Rathore
Department of Mechanical Engineering
Malaviya National Institute of
    Technology
Jaipur, Rajasthan, India

Bakhtawer Razaq
Department of Physics
University of Gujrat, Hafiz Hayat
    Campus
Gujrat City, Pakistan

Hamaela Razaq
Department of Chemical Engineering
Shanghai Jiao Tong University
Shanghai, China

Muhammad Rizwan
School of Physical Sciences
University of the Punjab
Lahore, Pakistan

Chetana Sabbanahalli
Department of Applied Sciences
Centre for Post Graduate Studies
Visvesvaraya Technological University
Bengaluru, Karnataka, India

Nanasaheb M. Shinde
School of Electrical and Engineering
Korea University
Seoul, Republic of Korea

Aleena Shoukat
Department of Physics
University of Gujrat, Hafiz Hayat Campus
Gujrat City, Pakistan

Anoop Singh
Department of Physics
University of Jammu
Jammu, Jammu and Kashmir, India

Nakshatra Bahadur Singh
Department of Chemistry and
    Biochemistry
Sharda University
Noida, India

Jhilmil Swapnalin
Multiferroic and Magnetic Material Research
    Laboratory (MMMRL)
Gandhi Institute of Technology and
    Management (GITAM) University
Bengaluru, Karnataka, India

Fateh Ullah
School of Engineering
Westlake University
Zhejiang, China

Ambreen Usman
Department of Physics
University of Gujrat, Hafiz Hayat Campus
Gujrat City, Pakistan

Vinay K. Verma
Department of Chemistry and Biochemistry
Sharda University
Noida, India

Qixun Xia
Henan Key Laboratory of Materials on
   Deep-Earth Engineering
School of Materials Science and Engineering
Henan Polytechnic University
Jiaozuo, China

Lele Zang
Henan Key Laboratory of Materials on
   Deep-Earth Engineering
School of Materials Science and Engineering
Henan Polytechnic University
Jiaozuo, China

1 Semiconductor Optical Fibers

Lele Zang, Qixun Xia, Zhijun Du, and Nanasaheb M. Shinde

CONTENTS

1.1 INTRODUCTION

With the change of times, semiconductor fibers have received great attention and developed rapidly. Initially, Kao and Hockham introduced the basic principle of the application of optical fiber in communication, and the bold idea was to use pure glass instead of copper wire to transmit signals through light [1]. Their work described the structure and material properties of insulating fibers for long-distance high-information optical communication, which opened the door to fiber optic communication, in everyday life. Optical fibers are used to transmit information over long distances because the loss of light in optical fibers is much less than the loss of electricity in wires.

Because the early commonly used communication window was 1310 and 1559 nm, the communication spectrum has been slow to develop, and the mid-infrared light became another choice for communication and sensing. Semiconductor core fiber has attracted much attention due to its special photoelectric properties. Bell Laboratories (the United States) successfully developed a semiconductor laser that can work continuously at room temperature. This is the world's first continuous wave gallium, aluminum or arsenide semiconductor laser operating at room temperature. The thin fiber is encased in a plastic sheath that bends but does not fold, which can easily break. We usually

DOI: 10.1201/9781003188582-1

choose to use a laser or light-emitting diode to send pulses to the fiber, while the fiber optic receiver uses a photosensitive element to detect the outgoing pulses. In 1977, Bell Research Institute and Nippon Telegraph and Telephone corporation almost simultaneously successfully developed a semiconductor laser with a life of 1 million hours (about 10 years in practice).

Later, semiconductor fiber was further developed, in 2006, the University of Southampton (SU) in the United Kingdom and Pennsylvania State University in the United States fabricated pure germanium core fiber using high-pressure chemical vapor deposition (HPCVD) [2–4]. First, they have measured two amorphous polycrystalline samples and their optical transport losses over a range of temperatures by a semiconductor deposition method using a silane/helium mixture flowing through a capillary pore [5]. Then Scott *et al.* used a powder-in-tube-type method to draw a silicon core fiber with a diameter of 10–100 μm and a length of about 7 cm at 1600°C [6, 7].

Based on the previous literature studies, different preparation methods have been used, i.e., molten core methods to synthesize high-efficiency semiconductor fibers. Clemson University (CU) in the United States in 2008, for the first time, used the molten core method to pull pure silicon, and then after a few years, a similar strategy was used to obtain germanium and indium antimonide core fiber. The semiconductor core melted into an optical fiber at optimized temperature (1000°C) allowed by the glass cladding (in this case, the glass is solid). However, the cladding glass acted as a crucible so that the molten semiconductor did not allow direct contact with the optical fiber [5].

At the same time, the fiber core can also be some new synthetic materials, and then traditional semiconductor materials can be replaced. The Massachusetts Institute of Technology (MIT) in the United States has studied mixtures of semiconductor and optical and electronic materials based on amorphous carbon, mainly focusing on the amorphous fibers composed of polymer semiconductors containing sulfur [6, 9]. In order to obtain semiconductor fibers with excellent optoelectronic characteristics and faster responsiveness, the research of optical semiconductor materials has entered the selenium era [8, 9]. This chapter mainly summarizes the structure, performance and related applications of the common mono and binary semiconductor core fibers to provide a good help for the future research.

1.2 SEMICONDUCTOR OPTICAL FIBER BACKGROUND

According to the principle of light propagation, the speed of light propagation is different in different media. When the light propagates in two different media, the light will be refracted and reflected at the interface of the two media. The angle of refraction of light with the angle of the incident light changes. When the angle of the incident light increases to a certain angle so that the angle of refraction reaches 90°, the refracted light will disappear completely, leaving only the reflected light, and the incident light will be reflected back. This is the principle of total reflection of light. Different materials refract light of the same wavelength at different angles, and the same substance refracts light of different wavelengths at different angles. Fiber optic communication is formed on the basis of the above principle [10].

Bare semiconductor optical fiber is generally divided into three parts: inside is a high-index semiconductor core, in the middle is a low-index silicon glass layer and the outer is reinforced with a resin layer. When the angle between the core layer and the outside surface is larger than the critical angle of total reflection, the light will reflect back after passing through the interface and continue to transmit forward in the core layer, and the cladding mainly plays a protective role. The electron microscopy image of the silica core and cladding layer is shown in Figure 1.1.

In 1870, John Tyndall introduced that light could be completely reflected along the boundary of a conducting medium, moving along the curve of the medium. This was the first experiment in transmitting light from a waveguide. By 1880, Bell was using sunlight as a light source, atmosphere as a medium and selenium crystals as a light receiver modulated by vibration. With a mere 213-m transmission range, this was the earliest wireless optical communication. However, optical communication has disadvantages: (1) there is no suitable light source. The directivity and correlation of

FIGURE 1.1 Displays an electron microscopy image of the cross section of the silica core and cladding layer. The core is roughly oval in shape. (From ref. [13]. Copyright 2008 OSA.)

ordinary light source (sunlight, light) is very poor, is close to noise and cannot adapt to the need of long-distance transmission; (2) there is no suitable transmission medium. With air as the transmission medium, the influence of optical signal is too great, and it must be guided through the low-loss dielectric device (waveguide). Therefore, a laser with stable frequency, good direction, good coherence and other advantages of advanced laser has been developed, which solves the problem of light sources being able to travel long distances in optical fibers. Subsequently, semiconductor core optical fiber has been widely concerned because of its special photoelectric properties. The advantages of using semiconductor fiber are also obvious: (1) near and middle infrared silicon and germanium core material is transparent with low loss; (2) semiconductor with high thermal conductivity and high optical damage threshold has good infrared and nonlinear characteristics (refractive index is large, the third-order nonlinear coefficient is large), semiconductor doping, thermal and photosensitive properties can be well applied to the sensing field.

The transmission of light in optical fiber can be interpreted as Equation (1.1): the light is incident from the medium with refractive index n_1 to the medium with refractive index n_2; there will be refraction on the medium interface [11]. That is:

$$\frac{\sin\varnothing_1}{\sin\varnothing_2} = \frac{n_2}{n_1} \tag{1.1}$$

where n_1 is the refractive index of the fiber core and n_2 is the refractive index of the cladding medium because n_1 is bigger than n_2, when the incident angle ϕ_1 is increased to a certain angle (ϕ_c), the refractive index will be equal to 90°, then the incident light will no longer enter the cladding medium, but the incident angle that begins to produce total reflection is called the critical angle, when ϕ_1 continues to increase, $\phi_2 > 90°$, total reflection occurs, the light then travels along the axis of the fiber. This is how the waveguide works. The light that does not satisfy the total reflection condition can only partially reflect at the interface, so some energy will radiate into the cladding, causing the light to travel at a very low efficiency. Usually, the light that can be repropagated in the waveguide is the transmission mode (guide mode), and the light that cannot be propagated is the guide mode.

For a given optical fiber structure and wavelength, the number of modes that can be propagated in an optical fiber is limited. The analysis shows that the propagation modes are:

$$M_{SI} = \left[\frac{V}{\left(\frac{\pi}{2} \right)} \right]^2 / 2 \tag{1.2}$$

Among them,

$$V = \frac{2\pi a}{\lambda} \sqrt{n_1{}^2 - n_2{}^2} \tag{1.3}$$

In Equation (1.3), V is called the normalized frequency or the nominal waveguide parameter, and $2a$ is the width of the optical waveguide type.

For a single-mode fiber with definite structure, there is no limitation on the wavelength of the base film. The wavelength corresponding to $V = 2.405$ is the cutoff wavelength of the higher-order mode, or the cutoff wavelength of the single-mode fiber. From the normalized frequency representation, the cutoff wavelength is easily found to be:

$$\lambda_c = \frac{2\pi a}{2.405} \sqrt{n_1{}^2 - n_2{}^2} \tag{1.4}$$

So, in this fiber, when the wavelength $\lambda > \lambda_c$, it will be a single-mode operation, and when $\lambda < \lambda_c$, it will be a multimode operation. It should be noted that since V is related to the wavelength of light, it is a single-mode fiber for a wavelength of light. For other light shorter than its λ_c, it is possible to propagate more than two modes and become a multimode fiber.

The relative refractive index difference is defined as the refractive index difference between the core and the cladding:

$$\Delta = \left[1 - \left(\frac{n_2}{n_1} \right)^2 \right] / 2 \tag{1.5}$$

When $\Delta < 0.01$, the upper equation is simplified to:

$$\Delta \approx \frac{n_1 - n_2}{n_1} \tag{1.6}$$

In the theory, the weak conductivity of light is one of the important differences between the light and the microwave dielectric waveguide. The basic meaning of weak conductivity is that a good optical fiber waveguide structure can be formed by a small difference of refractive index, and it is convenient for fabrication.

In numerical aperture (NA) of optical fiber, in order to characterize the excitation and coupling degree of incident light in optical fiber, the parameter of NA of optical fiber is often used. By optical convention, it is defined as:

$$NA = n_1 \sin \varnothing_c \tag{1.7}$$

In step fibers, for meridional rays (in a fiber, any plane passing through the central axis is called the meridional plane, and the rays that travel within the meridional plane are called the meridional rays):

$$NA = \sqrt{n_1{}^2 - n_2{}^2} \approx n_1 \sqrt{2\Delta} \tag{1.8}$$

For oblique light (light that is only unbalanced and does not intersect with the central axis of the fiber):

$$NA = \frac{\sqrt{n_1^2 - n_2^2}}{\cos\gamma} \tag{1.9}$$

In optical fiber coupling, the objective lens with NA to meet the requirement of NA of optical fiber is adopted, otherwise the loss will increase because of the light radiation in the continuous position. In the Gauss approximation, we can take $1/e^2$ as the NA.

1.3 SYNTHESIS APPROACHES

There are many methods to fabricate the optical fiber, including in-tube CVD or in-bar CVD (chemical vapor deposition), AVD (axial vapor deposition) and PCVD (plasma chemical vapor deposition). Either way, the prefabricated rod should be made at high temperature first, heated in a high-temperature furnace, it can then be softened and drawn into filaments, which are then coated and molded into fiber core wire. In the optical fiber manufacturing process, each process is controlled by the computer, so the calculation results require precision.

1.3.1 UNARY IV SEMICONDUCTOR CORE FIBER

For unary semiconductor optical fiber, silicon and germanium were selected as the main research objects. In a core melting method, the core material has to be allowed to melt at a temperature where the cladding glass is softened as well as draws into the optical fiber [12, 13]. The cladding of the silicon core fiber is the silicon glass, and the germanium core fiber is the borosilicate glass. Germanium core fibers are drawn at approximately 1000°C using a Heathway drawing tower for Clemson University, while silicon core fibers are also drawn at approximately 1950°C using a Heathway drawing tower for conventional telecommunication grade quartz fibers. The single semiconductor core drawing the operation process is shown in Figure 1.2.

At 1950°C, the melting point of the germanium core is much higher than that of the silicon core, and the melt is then coated by a viscous silicon coating. After the fiber straightens, generally in the diameter 1- to 2-mm scope, its center diameter is about 60 and 120 m, respectively. About 30 m of fiber was pulled out, but because of its relatively coarse diameter, the fiber was cut off by about 1.5 m [13]. At 1000°C, the Ge core was higher than its melting temperature (938°C), the liquid was surrounded by a softened glass cladding, and the fiber with a diameter of about 250 m and a core size of 15 μm [12]. Another advantage of silicon and germanium is that they can flow during the drawing process and form the shape of the glass coating, thus providing good continuity in length [5].

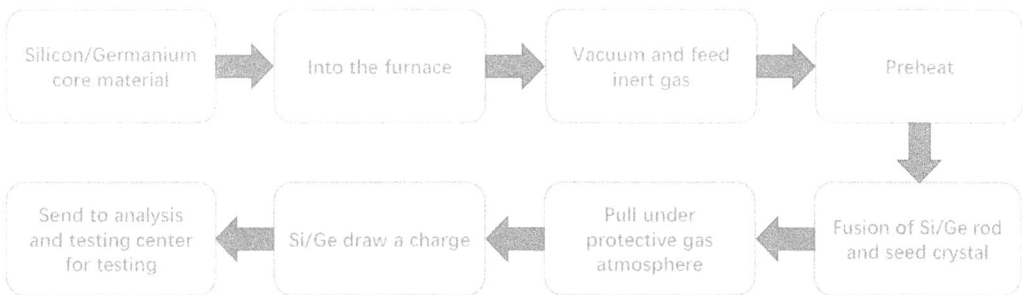

FIGURE 1.2 The flow chart of drawing operation for single semiconductor core.

1.3.2 BINARY III–V SEMICONDUCTOR CORE FIBER

Binary semiconductor fiber can integrate semiconductor materials with different characteristics into the same fiber, which can further improve the optical properties of semiconductor fibers and strengthen the integration degree of optoelectronic integrated devices. InP and InSb were selected as the main research objects. They manufactured a newfangled silica optical fiber mixed with nano-semiconductor material InP in cladding layers by adopting an ordinary modified chemical vapor deposition (MCVD) process.

InP powder was first placed at the entrance of the silicon substrate tube and diffused into the tube through high-temperature vaporization, thus completing the deposition process [14, 15]. With the flowing of nitrogen gas, after evaporation, InP was deposited on the silicon dioxide and finally on the inner surface, where it exists in the form of nanoparticles [16]. The InP nanoparticles were taken into the multi-hole soot of an MCVD silica preform before fiber was drawn [17]. Using the core method where the fusion-covering glass was selected, so that it can be used to draw at temperatures higher than the melting point of the core component, and the purpose of this operation was to confirm whether a binary semiconductor can also be subjected to this molten core method [17].

Clemson University mapped an InSb semiconductor core fiber clad in phosphate glass, which has been prepared at approximately 700°C using Heathway. As the indium antimonide melts, the liquid flows and pulled out at about 0.5 m/min. The molten indium antimonide fluid was acquired in the shape of a glass-coated tube. Since the sample was in small size, InSb chips were provided only about 2 m in diameter and 1 mm in fiber, a core size of about 230 μm [17]. We can easily choose a core ratio of about 5 or greater, which dramatically turns out to be appropriate so that there was an adequate quantity of softened cladding glass to physically maintain the weight of the molten core at the tensile temperature [16].

1.4 STRUCTURES AND OPTICAL PROPERTIES

1.4.1 STRUCTURES

Generally speaking, the optical fiber is composed of core and cladding, which transmits the signal, and the cladding has different refractive indices with the core, so the optical signal is transmitted and protected in the core.

Figure 1.3 shows the electron microscope images of Si and Ge crystalline core optical fibers. Optic fibers usually structured contain a semiconductor fiber core at the center of the light; Figure 1.3(a) shows a diameter of 5–75 μm, the role is to transmit light waves. The glass cladding is

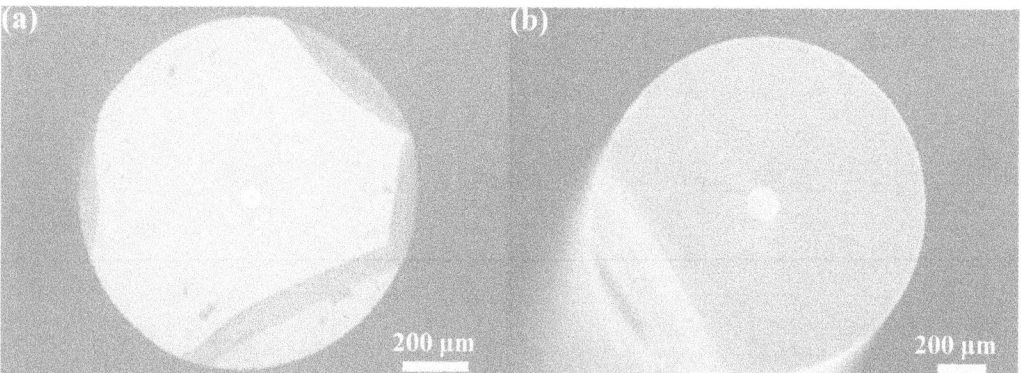

FIGURE 1.3 Displays the electron microscope images of crystalline core optical fibers: (a) Si and (b) Ge. (From ref. [5]. Copyright 2010 Elsevier Inc.)

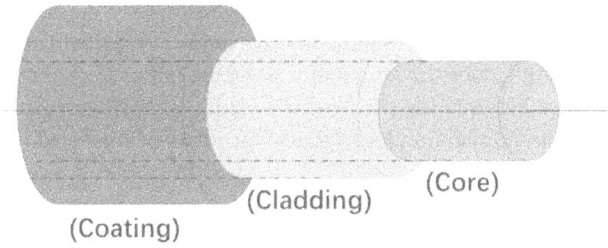

FIGURE 1.4 Simple structure of semiconductor fiber.

in the outer layer, while the core is in the inner layer with a diameter ranging from 100 to 150 μm, as shown in Figure 1.3(b).

The core and cladding make up the bare fiber, and the cladding is made of high-purity silicon dioxide (Figure 1.4). In order for light waves to travel through the core, materials should be doped differently or pure semiconductor elements should be used; the refractive index n_2 of the cladding material is smaller than that of the fiber material n_1, that is, the condition of the light conduction of the semiconductor fiber is $n_1 > n_2$. The external diameter of fiber is generally 125 μm, and for the internal diameter, most of the single-mode semiconductor fiber is 9 μm, and the multimode fiber is 50–62.5 μm.

1.4.2 OPTICAL PROPERTIES

1.4.2.1 Refractive Index

Refractive index is an important parameter for optical fiber. According to different refractive indices, the optical fibers can be roughly divided into three types, step-type optical fiber (SIF), graded index fiber (GIF) and W-type optical fiber (double-clad optical fiber) (WIF), as shown in Figure 1.5.

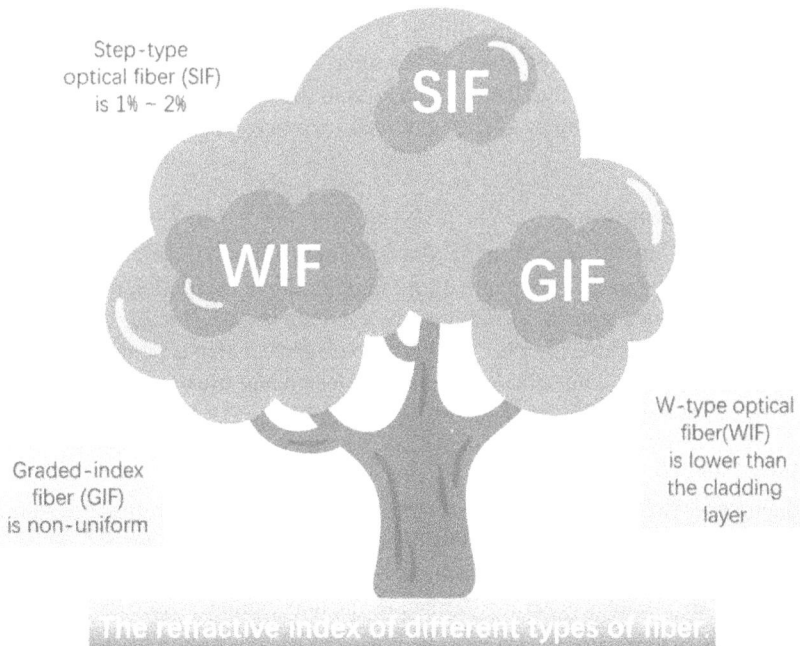

Step-type optical fiber (SIF) is 1% ~ 2%

SIF

WIF

GIF

Graded-index fiber (GIF) is non-uniform

W-type optical fiber(WIF) is lower than the cladding layer

The refractive index of different types of fiber

FIGURE 1.5 The refractive indices of various optical fibers.

SIF: the core refractive index is evenly distributed, and the relative refractive indices difference between core and cladding is 1%–2%. GIF: the refractive index of the fiber core is non-uniform, and it is maximum at the axis, but it decreases gradually along the radius in the cross section of the fiber and reaches at the interface between the core and the cladding. WIF: a buffer layer, the refractive index of which is lower than the cladding layer, is arranged between the cladding layer and the core making the cladding refractive index between the cladding layer and the core [18]. A dispersion-flattened fiber with a small dispersion variation between 1.3 and 1.6 μm or a dispersion-shifted fiber with zero dispersion wavelength shifted to 1.55 μm can be realized.

1.4.2.2 Dispersion
Dispersion is the coarsening of the bandwidth of a light pulse as it travels along a length of fiber. It is the main factor limiting the transmission rate. For inter-mode dispersion, it occurs only in multimode fibers because different modes of light travel different paths. For material dispersion, different wavelengths of light travel at different speeds [19]. The waveguide dispersion occurs because light energy travels at slightly different speeds through the core and cladding. In the single-mode fiber, it is very important to change the fiber dispersion by changing the inner structure of the fiber [20].

1.4.2.3 Attenuation
There are four main loss factors: the first is the absorption loss of core and cladding material; the second is the scattering loss of core and cladding material; the third is the scattering loss of the waveguide caused by the random distortion or roughness of the surface and the fourth is the radiation loss caused by the curved surface of the fiber. In general, intrinsic absorption and intrinsic scattering are inherent losses of optical fiber, while impurity absorption is non-inherent loss of optical fiber.

1.4.2.4 Numerical Aperture
The NA of the fiber represents the light-receiving ability of the fiber end face, and its value should take into account the light-receiving ability of the fiber and the influence on the mode of dispersion. In optics, NA is one of the parameters that indicates the performance of optical lens.

The greater the NA, the stronger the ability of the fiber to receive light. From the point of view of increasing the optical power entering the fiber, the larger the NA, the better the performance, because the larger NA of the fiber is beneficial to the docking of the fiber. However, when NA is too large, the mode distortion of the fiber will increase, which will affect the bandwidth of the fiber.

1.4.2.5 Cutoff Wavelength
The cutoff wavelength is also an important parameter of optical fiber. If the cutoff wavelength of the fiber is larger than the operating wavelength of the fiber, multiple oscillation modes will be formed in the fiber, and single-mode transmission cannot be carried out [21]. One of the main transmission characteristics of the single-mode fiber is the cutoff wavelength, which is of great significance to the manufacturers of optical fiber and cable as well as the design and use of the optical fiber transmission system by users of optical fiber and cable [22].

1.4.2.6 Mode Field Diameter
Mode field diameter (MFD) is used to characterize the distribution state of fundamental mode light in the core region of the single-mode fiber. The light intensity of the fundamental mode is the highest at the axial line in the core region and gradually decreases with the increase of the distance away from the axial line. The MFD is generally defined as the maximum distance between two points in each point where the light intensity is reduced to $1/(e^2)$ of the maximum light intensity at the axis line. The diameter of the mode field is related to the wavelength used. The diameter of the mode field increases with the increase of the wavelength.

1.4.2.7 Scattering

The scattering is the loss of light energy due to imperfections in the basic structure of light, when the transmission of light is no longer unidirectional.

1.5 THE APPLICATION OF SEMICONDUCTOR FIBER IN OPTICS

1.5.1 SEMICONDUCTOR AS OPTICAL AMPLIFIER

The amplification characteristic of semiconductor optical amplifier mainly depends on the dielectric characteristic of an active layer and the characteristic of laser cavity [23]. Although it is also a particle number inversion amplification luminescence, but the luminescent medium is non-equilibrium carrier, that is, electron hole pair rather than rare elements. Semiconductor luminescence can be divided into photoluminescence, electroluminescence and cathodoluminescence according to different excitation modes [24]. There are two types of semiconductor optical amplifiers (SOAs): the normal semiconductor laser used as an optical amplifier, called a F-P laser diode amplifier (FPA) and a F-P laser with anti-reflection film on both ends to eliminate the reflection at both ends to obtain broadband, high output and low noise [25]. The advantages of SOA are simple structure, small size, full use of existing semiconductor laser technology, mature fabrication process, low cost, long life, low power consumption and easy integration with other optical devices. In addition, its working band can cover the band of L. 3–1.6/mm, which cannot be achieved by EDFA or PDFA. In addition to optical amplification, SOA can be used as an optical switch and a wavelength converter.

At present, the effective control of its pulse width is carried out by using SOA. When the pulse signal is input into the semiconductor optical amplifier, the wavelength of the optical signal from the semiconductor optical amplifier will be narrowed to meet certain technical requirements, and the optical pulse signal through the semiconductor optical amplifier has a good symmetry [26]. Therefore, the semiconductor optical amplifier plays a role in the compression and shaping of optical pulse signals [27].

In conclusion, the semiconductor optical amplifier plays an important role in the field of communication, and an expansion is certainly expected in their application in the fiber optic communication in the future, so the research of new semiconductor optical fiber materials is an indispensable part in the field of communication [28].

1.5.2 SINGLE-CHIP SILICON PHOTONIC CRYSTAL FIBER PROBE SENSOR

Photonic crystal fiber (PCF) is a typical representative of two-dimensional photonic crystal. Compared with the traditional optical fiber, the silicon-air structure PCF is made up of tiny air holes arranged in order [29]. Since the first PCF was developed by Knight *et al.* in 1996, PCF has attracted much attention, and it has the advantages listed as non-cutoff single-mode characteristics, low-loss characteristics, flexible dispersion characteristics, controllable nonlinearity, strong birefringence effect, microstructural design modification, etc. The properties of PCF can be changed by changing the size and arrangement of air holes [29]. Optical fiber sensors have wide application prospects due to their obvious advantages such high sensitivity, anti-interference, simple structure, small volume, light weight, light path, bending property, the quality of the small impact, easy to form a network, etc. [30]. However, the optical fiber sensors using an ordinary optical fiber as the sensing element have some difficulties to overcome, such as large coupling loss, poor polarization retention and cross-sensitivity, the further improvement in the performance of the optical fiber sensors is limited [31–33].

The coupler sends the light to the PC end of the fiber. PC's reflected light is collected in OSAs. PCF has a novel structure and unique optical characteristics. With the development of fiber manufacturing technology, the characteristics of PCF will be developed and applied in sensing technology [34]. The future development of PCF sensors will focus on networking, integration, all-fiber and new sensing mechanism and scheme exploration.

1.5.3 SEMICONDUCTOR OPTICAL FIBER LASERS

Laser, through the mode selection of the cavity and the amplification of the gain medium, through the oscillation and amplification, achieved monochromatic, collimation, coherence of a very good beam. There are many types of lasers, but their necessary components are resonant cavity, gain medium and pump source [35].

The semiconductor laser works by excitation using semiconductor materials (that is, electrons) to jump between energy bands and emit light, and using semiconductor crystals to form two parallel mirror surfaces as mirrors to form a resonant cavity, make light to oscillate, feedback, produce light radiation amplification and output laser [35].

A semiconductor laser is a device that produces stimulated emission from a substance that works with a certain list of semiconductor materials. The principle is to achieve the inversion of the number of non-equilibrium carriers between the energy bands (conduction band and valence band) of semiconductor stuffs and the energy levels of journal (acceptor or donor); stimulated emission occurs when a large number of electrons in the inversion state are combined with holes.

The semiconductor laser is a big family of lasers. Compared to solid-state lasers, gas lasers and other types of lasers, it has the advantages of small volume, light weight, high electro-optic conversion efficiency, direct modulation and convenient use, that's why it works so well in fiber optic communication [35]. It has important applications in optical communication [36], optical transformation, optical interconnection, parallel optical wave system, optical information processing and optical storage. The advent of the semiconductor laser has greatly promoted the development of information optoelectronics technology [37]. Of course, it is also used in many other areas, such as: (1) application in military field, such as laser guidance and tracking, laser radar, laser communication power supply, laser simulation weapon, laser aiming alarm and laser gyro. At present, the world's developed countries are very concerned about the development of high-power semiconductor lasers and their military applications; (2) for printing and medical applications, such as CD players and DVD systems, especially red, green and blue surface-emitting lasers. Blue and green semiconductor lasers are used for laser printing, reading, writing and for screen color displays [38]; (3) extensive applications in optoelectronics, such as ultrahigh density, big area emission, optical reserving, lighting, optical signal, optical decoration, laser fabrication and medical treatment [35].

To sum up, fiber laser technology is a new technology research hotspot, which is being highly valued and rapidly developed, and it involves a wide range of scientific research and product application fields, and it has huge potential application value and broad market prospects.

1.6 CONCLUSION

The 21st century is an era of information explosion, people's demand for information is more and more extensive. Therefore, in recent years, scientists have gradually deepened their research on semiconductor fiber and made some breakthrough achievements.

In this chapter, the basic characteristics of semiconductor core fiber are presented by introducing mono and binary semiconductor fiber materials. In the fabrication of semiconductor core fiber, we mainly discuss the two methods of high-pressure microfluidic chemical deposition and core fusion. As a new material of optical fiber, it is not difficult to see the advantages of semiconductor materials, such as large refractive index, large third-order nonlinear coefficient, 3–4 orders of magnitude of Raman gain coefficient compared with quartz glass and transparent waveguide transmission in some semiconductors. But similarly, the performance of the semiconductor core also needs to be improved, such as its large loss, oxygen and sediment content and material equipment are not perfect and so on.

Even with these disadvantages, the developments of semiconductor fibers in amplifier, sensor and laser direction should not be underestimated. It is hoped that after shortcomings, the development of semiconductor core fiber will take a qualitative leap in the future.

ACKNOWLEDGMENTS

This study was supported by the China Postdoctoral Science Foundation (2019M652537), the Fundamental Research Funds for the Universities of Henan Province (NSFRF210326), the Fundamental Research Funds for the Universities of Henan Province (NSFRF210326), the Henan Postdoctoral Foundation (19030065), the Henan Province Key Science and Technology Research Projects (202102310628), the Foundation of Henan Educational Committee (20B430006) and the Doctoral Foundation of Henan Polytechnic University (B2019-41).

REFERENCES

1. C. K. Kao, A. G. Hockham, Dielectric-fibre surface waveguides for optical frequencies, Optoelectronics IEE Proceedings Journal, 133 (1986) 441–444.
2. B. R. Jackson, P. J. A. Sazio, J. V. Badding, Single-crystal semiconductor wires integrated into micro-structured optical fibers, Advanced Materials, 20 (2007) 1135–1140.
3. Sazio P. J. A., Microstructured optical fibers as high-pressure microfluidic reactors, Science, 311 (2006)1583–1586.
4. D. J. Won, M. O. Ramirez, H. Kang, V. Gopalan, N. F. Baril, J. Calkins, J. V. Badding, P. J. A. Sazio, All-optical modulation of laser light in amorphous silicon-filled microstructured optical fibers, Applied Physics Letters, 91 (2007) 615–646.
5. J. Ballato, T. Hawkins, P. Foy, B. Yazgan-Kokuoz, C. Mcmillen, L. Burka, S. Morris, R. Stolen, R. Rice, Advancements in semiconductor core optical fiber, Optical Fiber Technology, 16 (2010) 399–408.
6. B. Scott, Fabrication of silicon optical fiber, Optical Engineering, 48 (2009) 100501.
7. B. L. Scott, W. Ke, G. Pickrell, Fabrication of n-type silicon optical fibers, IEEE Photonics Technology Letters, 21 (2009) 1798–1800.
8. M. Bayindir, A. F. Abouraddy, O. Shapira, J. Viens, D. S. Saygin-Hinczewski, F. Sorin, J. Arnold, J. D. Joannopoulos, Y. Fink, Kilometer-long ordered nanophotonic devices by preform-to-fiber fabrication, IEEE Journal of Selected Topics in Quantum Electronics, 12 (2006) 1202–1213.
9. D. S. Deng, N. D. Orf, S. Danto, A. F. Abouraddy, J. D. Joannopoulos, Y. Fink, Processing and properties of centimeter-long, in-fiber, crystalline-selenium filaments, Applied Physics Letters, 96 (2010) 023102.
10. G. P. Agrawal, Fiber-optic communication systems, Nasa STI/Recon Technical Report A, 93 (2002) 12–20.
11. K. Iizuka, Fiber optical communication, Springer Series in Optical Sciences, 35 (2008) 333–369.
12. J. Ballato, T. Hawkins, P. Foy, B. Yazgan-Kokuoz, R. Stolen, B. Jalali, R. Rice, Glass-clad single-crystal germanium optical fiber, Optics Express, 17 (2009) 8029–8035.
13. T. Hawkins, P. Foy, J. Ballato, R. Stolen, B. Kokuoz, M. Ellison, C. Mcmillen, J. Reppert, A. M. Rao, M. Daw, Silicon optical fiber, Optics Express, 16 (2008) 18675.
14. J. Ballato, T. Hawkins, P. Foy, B. Kokuoz, R. Stolen, C. Mcmillen, M. Daw, Z. Su, T. M. Tritt, M. Dubinskii, On the fabrication of all-glass optical fibers from crystals, Journal of Applied Physics, 105 (2009) 053110.
15. J. Ballato, E. Snitzer, Fabrication of fibers with high rare-earth concentrations for Faraday isolator applications, Applied Optics, 34 (1995) 6848–6855.
16. F. Pang, X. Zeng, Z. Chen, T. Wang, Fabrication and characteristics of silica optical fiber doped with InP nano-semiconductor material, Optical & Quantum Electronics, 39 (2007) 975–981.
17. J. Ballato, T. Hawkins, P. Foy, C. Mcmillen, R. R. Rice, Binary III-V semiconductor core optical fiber, Optics Express, 18 (2010) 4972–4979.
18. M. Anani, C. Mathieu, S. Lebid, Y. Amar, Z. Chama, H. Abid, Model for calculating the refractive index of a III–V semiconductor, Computational Materials Science, 41 (2008) 570–575.
19. T. L. Tansley, D. F. Neely, C. P. Foley, Optical dispersion in zinc oxide, Physica Status Solidi (A), 77 (1983) 491–496.
20. T. Berceli, E. Udvary, A. Hilt, A new equalization method for dispersion effects in optical links, Anniversary International Conference on Transparent Optical Networks, (2008) 98–101.
21. P. Cochrane, M. Brain, Future optical fiber transmission technology and networks, Communications Magazine, IEEE, 26 (1988) 45–60.
22. D. Keller, Optical fiber cable and device for manufacturing a cable of this kind, US, US 5621842 A [P], (1997).

23. K. Obermann, S. Kindt, Noise characteristics of semiconductor-optical amplifiers used for wavelength conversion via cross-gain and cross-phase modulation, IEEE Photonics Technology Letters, 9 (1997) 312–314.

24. G. Hunziker, R. Paiella, K. J. Vahala, U. Koren, Measurement of the stimulated carrier lifetime in semiconductor optical amplifiers by four-wave mixing of polarized ASE noise, Photonics Technology Letters IEEE, 9 (2002) 907–909.

25. K. K. Qureshi, H. Y. Tam, Multiwavelength fiber ring laser using a gain clamped semiconductor optical amplifier, Optics & Laser Technology, 44 (2012) 1646–1648.

26. F. R. Barbosa, C. Coral, J. R. Caumo, A. Flacker, Hermetically packaged semiconductor optical amplifier for application In singlemode fiber systems, SBMO International Microwave Conference/Brazil, 1 (1993) 129–134.

27. A. E. Kelly, Ultra high-speed wavelength conversion and regeneration using semiconductor optical amplifiers, Optical Fiber Communication Conference & Exhibit, (2001) MB1.

28. S. M. Idrus, H. K. Lim, H. Y. Looi, Structure and Characteristic of the Semiconductor Optical Amplifier in Optical Fiber Communication System, Penerbit UTM, (2008).

29. W. I. Jung, B. Park, J. Provine, T. R. Howe, O. Solgaard, Highly sensitive monolithic silicon photonic crystal fiber tip sensor for simultaneous measurement of refractive index and temperature, Journal of Lightwave Technology, 29 (2011) 1367–1374.

30. A. Wang, Y. Zhu, G. Pickrell, Optical fiber high-temperature sensors, Optics and Photonics News, 20 (2009) 27–31.

31. W. Liang, Y. Huang, Y. Xu, R. K. Lee, A. Yariv, Highly sensitive fiber Bragg grating refractive index sensors, Applied Physics Letters, 86 (2005) 647–688.

32. G. Meltz, W. W. Morey, W. H. Glenn, Formation of Bragg gratings in optical fibers by a transverse holographic method, Optics Letters, 14 (1989) 823–825.

33. T. Erdogan, V. Mizrahi, P. J. Lemaire, D. Monroe, Decay of ultraviolet-induced fiber Bragg gratings, Journal of Applied Physics, 76 (1994) 73–80.

34. I. W. Jung, B. Park, J. Provine, R. T. Howe, O. Solgaard, Photonic crystal fiber tip sensor for precision temperature sensing, IEEE Leos Meeting Conference, (2009) 761–762.

35. M. R. Stiglitz, C. Blanchard, Of masers, lasers and optical fiber, Microwave Journal, 34 (1991) 5287077.

36. W. I. Way, F. K. Tong, A. E. Willner, Optical Fiber Communication, McGraw-Hill, (1983).

37. D. Dopheide, H. Többen, V. Strunck, G. Grosche, H. Müller, Realization of high performance LDA-systems using optical amplifiers, high power semiconductor lasers and optical fiber lasers, Journal of Visualization, 2 (2000) 281–292.

38. X. Wang, Surface Emitting Semiconductor Laser Thermal and Radiation Analysis, Ms Electrical Engineering, (2013).

2 Optical Properties of Semiconducting Materials for Solar Photocatalysis

Dhirendra Singh Rathore and Sapana Jadoun

CONTENTS

2.1 INTRODUCTION

Water contamination has been of great concern due to the increment of industrial discharges, excessive uses of fertilizers, pesticides as well as domestic waste. For the treatment of wastewater, many processes such as sterilization, flocculation, filtration, and chemical oxidation of organic pollutants are currently in use (Bashir et al. 2020). Biological treatment is followed after the elimination of particles in suspension but all the contaminants are not biodegradable and these non-biodegradable contaminants are called bio-recalcitrant organic compounds (BROC). To eliminate these BROC, some advanced oxidation technologies (AOT) are employed in which free hydroxyl radicals ($\cdot$OH) are formed, which possess rich oxidizing power compared to other oxidants (Robert and Malato 2002).

Semiconductor materials have attracted special attention due to their outstanding properties and potential applications in solar photocatalysis for the degradation of environmental pollutants (Zhou et al. 2018). These materials possessing high photosensitivity, wide bandgap, and non-toxic nature act as promising photocatalysts. The optical properties of these semiconductors have been widely studied for solar photocatalysis for ages (Halder et al. 2020). These semiconductors, including TiO_2 (Nakata and Fujishima 2012),

DOI: 10.1201/9781003188582-2

13

ZnO (Jadoun et al. 2020), CuS (Basu, Garg, and Ganguli 2014), CdSe (Elmalem et al. 2008), CdS (Liu et al. 2019), and MoS_2 (Li, Meng, and Zhang 2018) (Shen et al. 2011) are n- and p-type semiconductors used as per the convenience of applications in various fields. The performance of photocatalytic materials depends on the optical properties of semiconductors such as absorption of light in the UV or visible region, narrow or wide bandgap, and tuning the surface morphology. There is a direct relationship between the optical properties and photocatalytic activities of a catalyst. Various defects or impurities also played a key role in tuning these optical properties for the irradiation under solar light (Pelaez et al. 2012).

2.2 SOLAR PHOTOCATALYSIS

Solar energy is an abundant, nonpolluting, inexpensive way to generate electricity as well as useful in wastewater treatment for environmental remediation and artificial photosynthesis (Kalogirou 2013). For this process, the term photocatalysis is used in which natural light can be utilized by some semiconductor materials known as photocatalysts. Due to the intrinsic nature and suitable electronic structure, semiconductor materials are the perfect candidate for solar photocatalysis (Kubacka, Fernandez-Garcia, and Colon 2012).

A semiconducting material that can be activated chemically by radiation is known as a heterogeneous photocatalyst. If it is irradiated by a light equal to and greater than its bandgap energy, an electron is excited from the valence band (VB) to the conduction band (CB) and leaves a hole in the VB. The electrons in the CB and holes in the valance band are responsible for the redox reaction at the surface of the catalyst. CB electrons are powerful reductants, while VB holes are oxidants. In the absence of an electron and hole scavenger, this charge carrier tends to recombine swiftly. Hence, the whole photocatalytic efficiency depends on the charge transfer to the surface and the recombination of electrons and holes. The recombination can be stopped by a suitable scavenger or some modifications in the surface structure of semiconducting material to trap the electrons and holes. The photocatalytic reactions produce the mineralized products during the reaction. Holes present in the system allow the one-electron oxidation step to yield hydroxyl radicals and these are responsible for the oxidation of organic contaminants to the gas or liquid phase. In the presence of air, oxygen can also behave as an acceptor of electrons and can be reduced to form a superoxide ion by the electrons present in the CB. Superoxide ion is also a strong oxidant, Figure 2.1 (Al-Rasheed 2005; Bard 1979; Colmenares and Xu 2016).

$$h^+ + H_2O \rightarrow HO^\bullet + H^+$$

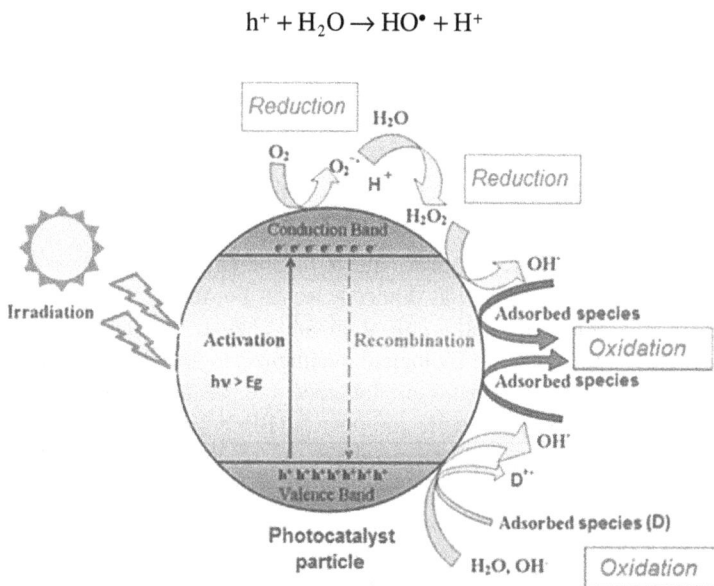

FIGURE 2.1 Mechanism of solar photocatalysis of heterogeneous catalyst. (Reprinted with permission from Spasiano et al. (2015).)

The bandgap width between the VB and CB shows the chemical bond's strength. Semiconductors such as metal oxides remain located at top of VB, i.e., +3 eV NHE (normal hydrogen electrode) or lower, and absorb energy under visible light (more than 400 nm) fall under solar photocatalysis (Spasiano et al. 2015).

2.3   SEMICONDUCTING MATERIALS

Semiconductors are the widely used photocatalysts for the decomposition and degradation of pollutants due to their structural properties that fulfill the photocatalysis requirements. The structural arrangements, generation of charge carriers, appropriate bandgap, reusability, capabilities of light absorption, high surface area, and stability are the main features that make semiconductors an outstanding material for photocatalysis (Makuła, Pacia, and Macyk 2018). These semiconductors include TiO_2, ZnO, CdS, CdSe, and SnO. (Jadoun, Verma, and Arif 2020; Verma, Arif, and Jadoun 2020). These materials absorb light, which activates these, and the electrons excited from the VB to the CB induce the formation of electron and hole pairs, which involves the redox reaction by absorbing O_2, H_2O, etc., on the surface of the photocatalyst. These generate the hydroxyl and superoxide reactive oxygen species, which interact with the pollutants and contaminants to reduce or degrade them into less harmful products or completely destroy them (Schneider et al. 2014). The main problems involved in photocatalysis due to these semiconductors are wide bandgap, absorption mainly in the UV region only, and low quantum yield due to less lifetime of charge carriers because of recombination of electrons and holes. These all factors become the hurdle in the photocatalytic activity of semiconductors to degrade pollutants (Karthikeyan et al. 2020). To overcome these problems and irradiation in the solar light, these semiconductors are combined with various other materials known as hybrid semiconducting materials as discussed under Section 2.5.

2.4   SYNTHESIS OF SEMICONDUCTING MATERIALS

Synthesis of these semiconducting materials has been done by various techniques and the method adoption depends on the availability of material and its type. Some common methods of synthesis of semiconducting materials are depicted in Figure 2.2 and are discussed below.

2.4.1   ARC DISCHARGE

The arc method includes an arc in an inert environment in between the two electrodes, which results in the nanomaterial's deposition on the electrode surface. This method is usually applied for the formation of CNT (carbon nanotubes) and fullerene as shown in Figure 2.2(a) (Seeger, Kohler-Redlich, and Ruehle 2000).

2.4.2   SONOCHEMICAL

This synthetic method includes the interaction via "acoustic cavitation" in between the ultrasound wave and precursor, which leads to the creation of bubbles and collects ultrasonic energy to oscillate and raise in size, as shown in Figure 2.2(b). The size development helps in collapsing and releasing the energy stored leading to the generation of nanoparticles. This method is used for the synthesis of graphene-Fe-TiO_2, Cl-co-doped TiO_2, etc. (Abdel Rahman et al. 2018).

2.4.3   EXFOLIATION

The exfoliation method can be used by two types (i) mechanical exfoliation and (ii) chemical exfoliation. Graphene from graphite nanomaterials is synthesized by mechanical exfoliation, as shown in Figure 2.2(c). This can be achieved with the help of scotch tape from the substrate peeling. Chemical exfoliation includes the peeling in the presence of chemicals via oxidation under stirring or ultrasonication,

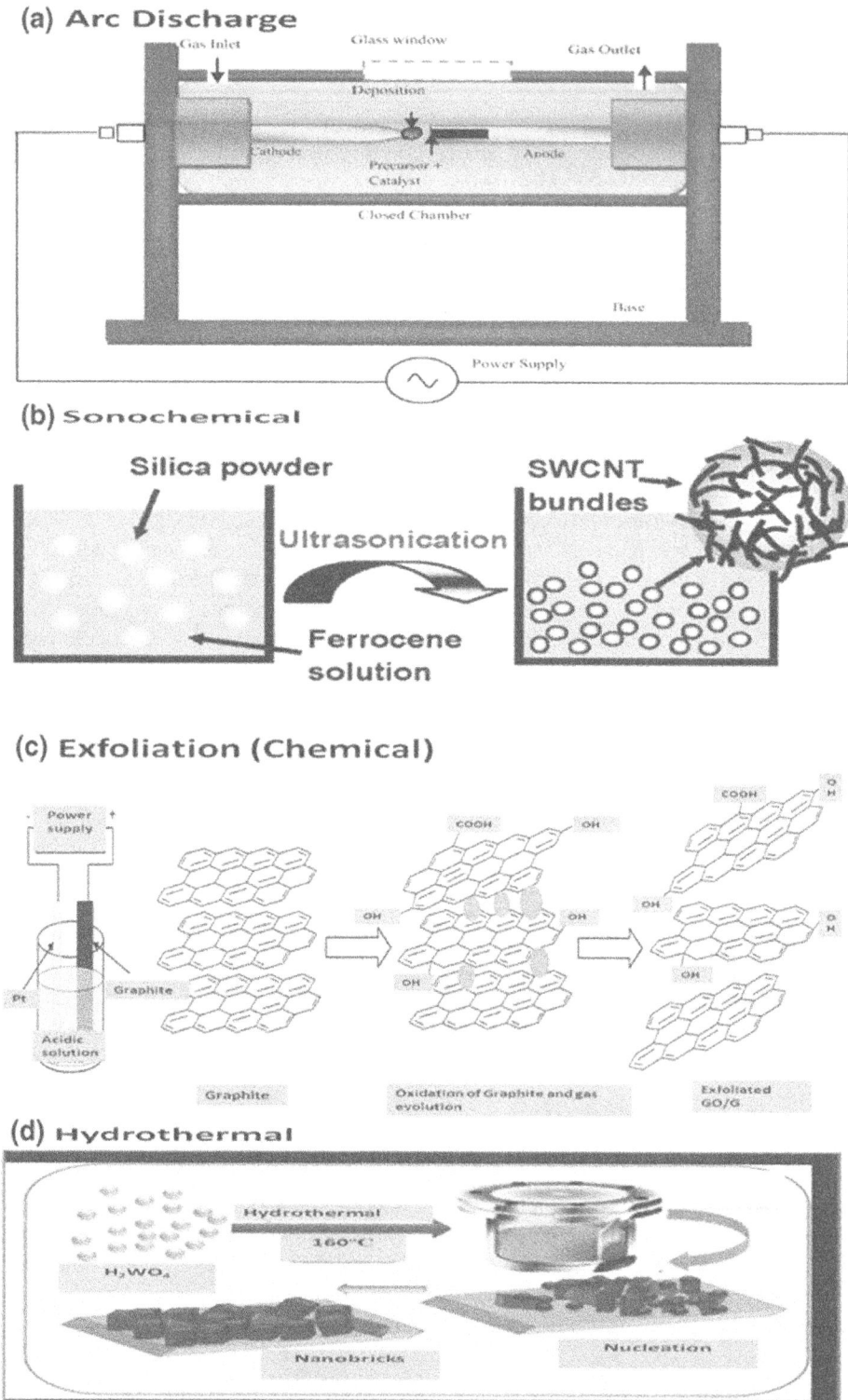

FIGURE 2.2 Schematic representation of various synthetic methods of nanostructures for energy harvesting (a) arc discharge, (b) sonochemical, (c) exfoliation, (d) hydrothermal. (Reprinted with permission from Rani et al. (2018).) (*Continued*)

(e) Solvothermal

(f) Microwave Assisted Synthesis

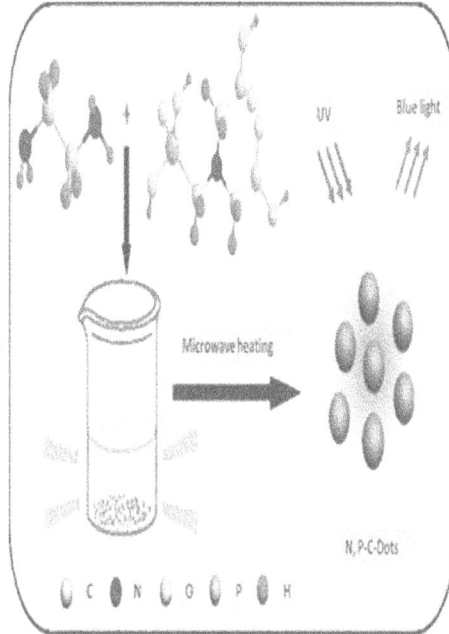

(g) Polyol Synthesis

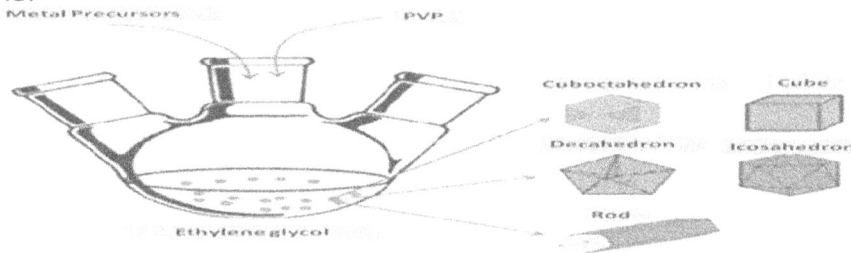

(h) CVD

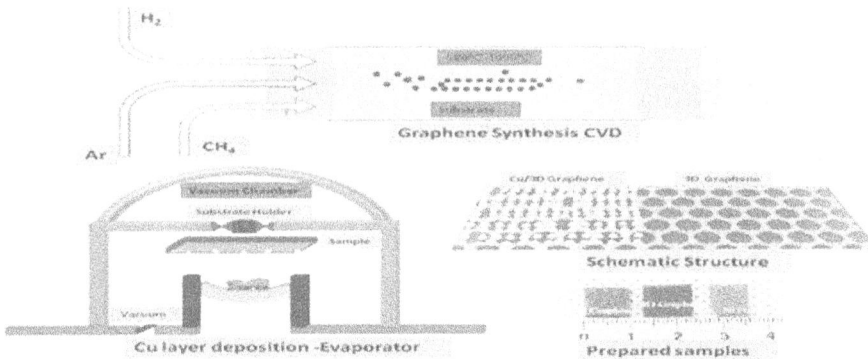

FIGURE 2.2 (*Continued*) Schematic representation of various synthetic methods of nanostructures for energy harvesting (e) solvothemral, (f) microwave-assisted, (g) polyol, (h) CVD. (Reprinted with permission from Rani et al. (2018).) (*Continued*)

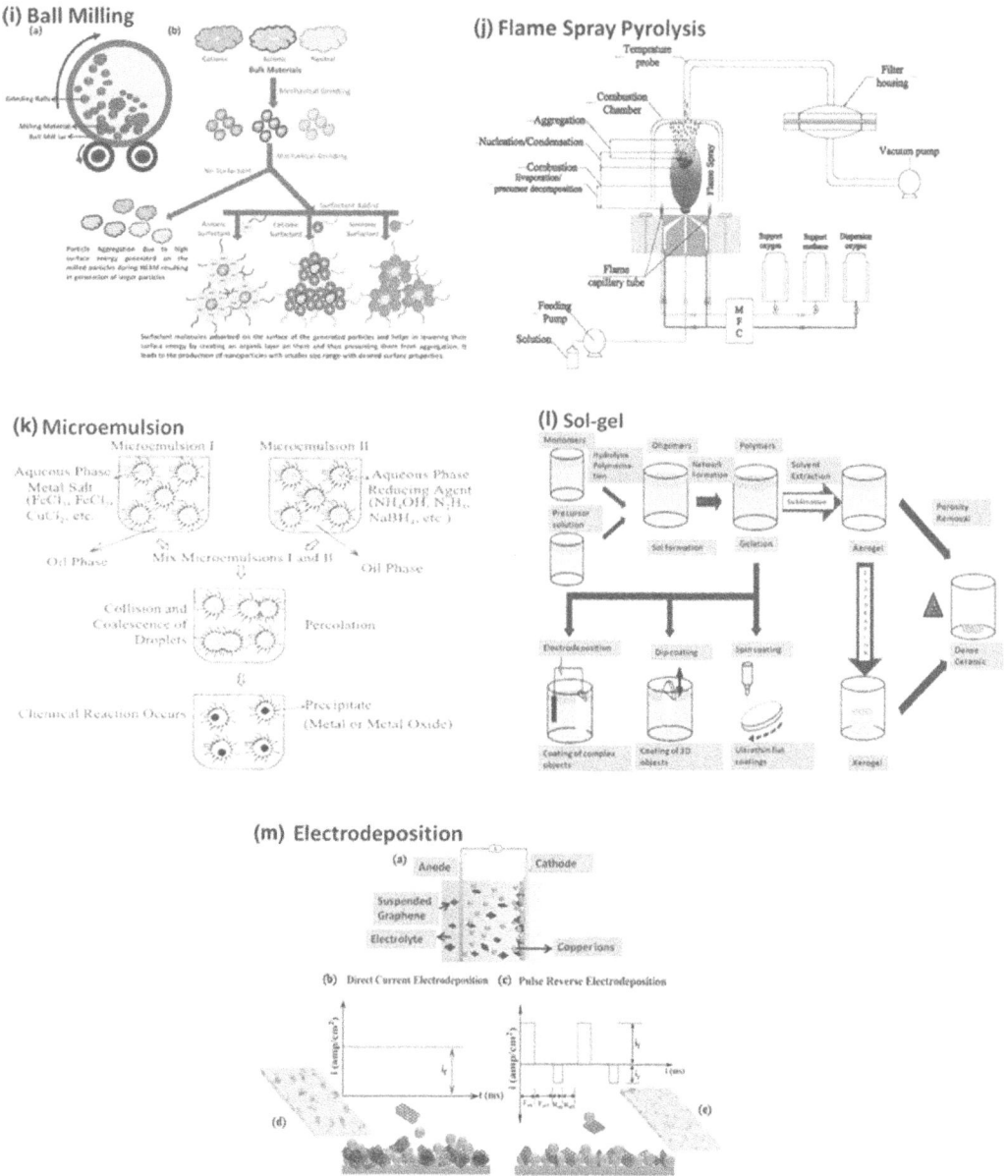

FIGURE 2.2 (*Continued*) Schematic representation of various synthetic methods of nanostructures for energy harvesting (i) ball milling, (j) flame spray pyrolysis, (k) microemulsion, (l) sol-gel, and (m) electrodeposition. (Reprinted with permission from Rani et al. (2018).)

which deteriorates the bonding of the material to produce nanomaterials. Following this method, some additional photocatalysts can also be synthesized along with graphite oxide (Liu et al. 2018).

2.4.4 HYDROTHERMAL

This method involves the heating of precursors in the autoclave within the temperature range 100–300°C at high pressure. This method comprises the steps of nucleation and growth of the crystal. The elevated temperature is responsible for generating high pressure and higher solubility of substrates. This is a very easy method of synthesis for several nanomaterials, as shown in Figure 2.2(d) (Zou et al. 2006).

2.4.5 SOLVOTHERMAL

This method differs from the hydrothermal method. The various conditions such as pressure, temperature, and use of solvent affect the reaction process. Sometimes, the size of nanomaterials can be decreased by increasing the temperature. This method is suitable for the synthesis of photocatalyst materials of defined morphology, as shown in Figure 2.2(e) (Tang et al. 2003).

2.4.6 MICROWAVE-ASSISTED SYNTHESIS

This method comprising the combination of high pressure and temperature improved the rate of reaction in the microwave due to rapid homogenous mixing of reactants. By combining this method with others, better yield can be achieved. Nanomaterials can be synthesized by this method with unique properties, as shown in Figure 2.2(f) (Hasanpoor, Aliofkhazraei, and Delavari 2015).

2.4.7 POLYOL SYNTHESIS

This method involves the boiling of polyol at high temperatures with capping agents and precursors. This is a liquid-phase method of synthesis, which includes solvents such as glycerol, ethylene glycol, butanediol, and diethylene glycol. As a precursor, this method includes halides, nitrates, sulfates, and metal salts, as shown in Figure 2.2(g) (Dong, Chen, and Feldmann 2015).

2.4.8 CHEMICAL VAPOR DEPOSITION (CVD)

It is a technique in which the surface of the substrate is coated with the nanomaterial of choice. The deposition or coating is enabled by the action of heat, which is further accomplished by the chemical reaction with antecedent gases. These reactions start with the reactants in reactors followed by the diffusion of reactant gas on the surface of the substrate. After that, gaseous molecules adsorb on the surface of the substrate and the surface reaction occurs, and lastly, desorption and removal of by-products occur. This technique is generally used for the synthesis of cellulose nanofiber (CNF), CNT, and metal oxides with tuned morphology, as shown in Figure 2.2(h) (Cong et al. 2014; Rashid et al. 2015).

2.4.9 BALL MILLING

This method involves the mechanical treatment of powdered samples to yield smaller homogenous nanoparticles. The charging of samples occurs with small balls present in the milling chamber to accomplish the grinding operation. Due to the rotation of the chamber, breaking of samples into small sizes via collision occurs. Selective nanomaterials can be synthesized by this technique, as shown in Figure 2.2(i) (Elilarassi and Chandrasekaran 2012).

2.4.10 FLAME SPRAY PYROLYSIS

This method based on the vapor phase operates at high temperatures. In this technique, the formation of high-energy flame occurs via a laser in the closed chamber by using carbon dioxide. The precursor in the form of an aerosol passed into the reaction space for the flame treatment. The swift cooling of the obtained product takes place for obtaining desired products. Hence, the factor that can dominantly affect this method is temperature, as shown in Figure 2.2(j). This technique is generally used for the synthesis of high surface area heterojunction composites and crystal-modified copper sulfide (CuS) (Chiarello, Selli, and Forni 2008).

2.4.11 MICROEMULSION

A microemulsion of water-in-oil or oil-in-water collides to nucleate and generate the nanoparticles in this method. It is also a liquid-phase synthesis method like polyol synthesis that comprises reverse micelle as reaction unit. This method offers the desired shape, size, and porosity to nanomaterials, as shown in Figure 2.2(k) (Tolia, Chakraborty, and Murthy 2012).

2.4.12 SOL-GEL TECHNIQUE

For the synthesis of high-purity materials, the sol-gel method is applied at room temperature. It adopts colloidal with transitional solution and gel-phase formation. This method includes the use of precursors in the form of metal alkoxides that assist the monomer in the polymerization reaction. Later on, to obtain desired gel, hydrolysis occurs followed by condensation and crosslinking. This technique is majorly used to synthesize composite materials and ceramics having a high surface area, as shown in Figure 2.2(l) (Tsay et al. 2010).

2.4.13 ELECTRODEPOSITION

Electrodeposition is a surface technique employed at room temperature in which the coating of the surface of the substrate can be done with nanomaterials by using the current to enhance the characteristics of nanomaterials. This method includes a cathode and an anode, which remain in an electrolyte solution, and on passing the electricity, reduction of cation takes place and deposition of nanomaterials occurs at the surface of the cathode as a thin layer, as shown in Figure 2.2(m) (Lincot 2005).

2.5 HYBRID SEMICONDUCTING MATERIALS

Hybrid semiconducting materials hold unique functionalities and optical properties, which makes them suitable for photocatalytic applications under visible light as well as in UV irradiation. Hybrid semiconductors provide flexibility to materials such as surface morphology and tuned bandgap via absorption in the desired field (Liao et al. 2019). These hybrid semiconducting materials can be divided into four groups based on the composition of materials as follows:

1. Inorganic semiconductor-insulator
   For various applications in sensors and electronics, these types of hybrid materials have been widely used. In these systems, the optical properties are generally dominated by semiconductors because insulators possess wide bandgap, while semiconductors possess small; hence, the absorption occurs in the visible region due to the small bandgap of semiconductors. If the absorption has to be considered in the UV region, the insulating properties can also be different. Hereafter, the insulator and semiconductor's dominance depend on the spectroscopic region selected, i.e., the band positions and edges alignment, for example, silica-coated CdSe/ZnS core-shell quantum dots (Gerion et al. 2001).
2. Inorganic semiconductor-semiconductor
   This type of system exhibits behavior similar to the inorganic semiconductor-insulator system but the only difference is another semiconductor's bandgap that is small as compared to the insulator. The other semiconductor can be inorganic (CdSe/ZnS) (Chauvire et al. 2015) or organic (PANI/TiO_2) (Jangid et al. 2020). In an inorganic semiconductor-organic semiconductor system, conjugated polymers play a key role by extended conjugation, which diminished the energy gap between HOMO (highest occupied molecular orbital) and LUMO (lowest unoccupied molecular orbital) and shifted the absorption to the visible region, and thus increased the chances to absorb solar light. The energy level of these is close to inorganic semiconductors (Suresh et al. 2008; Chou et al. 2006).

3. Inorganic semiconductor-metal

Both, inorganic semiconductor and metal are electronically and optically active having the absorption in the visible region due to their low-energy electronic transition between 1 and 3 eV. These hybrids are interesting materials due to their varied optical properties and interactions. These are more complex hybrids compared to the above-discussed systems under (i) and (ii). The absorption spectrum is generally the additive of both the material's spectra and the photoluminescence properties such as increment and decrement of intensity depending on the interaction between the two components because generally metals possess very low photoluminescence (Shi et al. 2021; Usman, Mendiratta, and Lu 2017).

4. Doped semiconductor nanomaterial

This is the special case of semiconductor hybrid materials. The properties and functionality of any semiconductor can be modified by doping for use in sensing, imaging, laser, light energy conversion, etc. The primary semiconductors are denoted as host semiconductor A, while the other dopant is denoted as B, which is used to alter the properties of the semiconductor. In the comparison of above-undoped systems, this type of material possesses extraordinary properties due to tuning of magnetic, electrical, and optical properties by doping (Chen et al. 2005; Liu and Swihart 2014).

2.6 OPTICAL PROPERTIES

The bandgap (eV) is the energy difference between the top of VB and the bottom of CB. This is the energy that is required. In semiconductors, this gap remains smaller as compared to insulators but wider compared to conductors. This is the minimum energy required to excite an electron from the filled VB to empty CB (Chan and Ceder 2010). The bandgap values of semiconductors with their edge potentials at zero (0) pH vs NHE are shown in Figure 2.3.

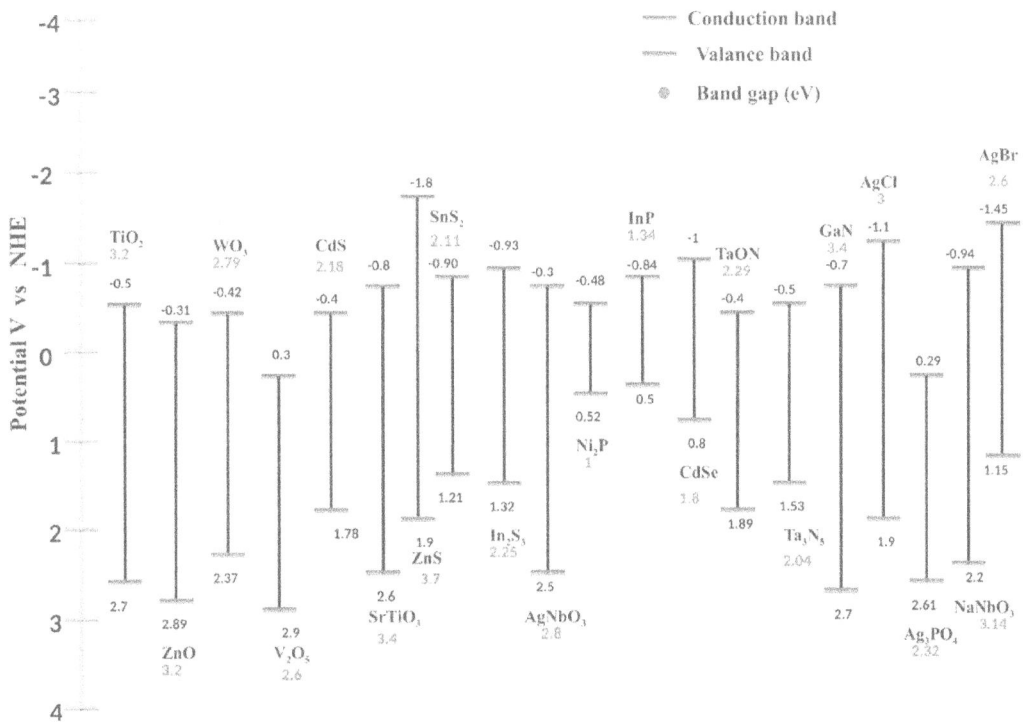

FIGURE 2.3 Representation of bandgap (Eg) values (in eV) and the edge potential of CB and VB at 0 pH vs NHE. (Reprinted with permission from Rani et al. (2018).)

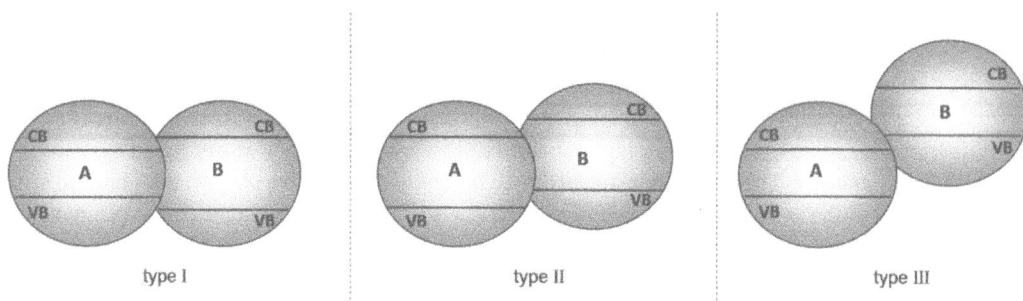

FIGURE 2.4 Various types of semiconductor heterojunctions. (Reprinted with permission from Marschall (2014).)

When the photocatalyst is designed using two distinct materials having different bandgaps, the band positions of both defined three types of heterojunctions as shown in Figure 2.4. In the type I heterojunction, it can be noticed that semiconductor A has a higher VB and lower CB than semiconductor B. In this type of heterojunction, the transfer of electrons occurs after sensitization from CB of semiconductor B to CB of semiconductor A while the holes are also transferred from semiconductor B to semiconductor A. Such type of arrangement favored the accumulation of all the charges on the surface of semiconductor A and recombination of charge occurs; hence, separation of charge carrier is minimal and diminishes the photocatalytic activity. The bandgap position of both the semiconductors in heterojunction type II is optimum for the separation of charge carriers. The photoexcited electrons migrated from CB of semiconductor B to CB of semiconductor A, while holes are transferred from VB of semiconductor A to B. Therefore, the holes and electrons are separated from each other by synergistic interaction. This type of heterojunction includes the composites for photocatalysis. Type III heterojunction possesses the bandgap positions in such a way that VB and CB of semiconductor A lie at lower positions than semiconductor B but the charge transfer in these heterojunctions occurs like heterojunction II. This type of arrangement of semiconductors is known as broken-gap situations.

To narrow the bandgap, the photocatalyst is needed to absorb light in the visible region to enhance the quantum efficiency, and to achieve this, some structural modifications are required. Metal or nonmetal doping is one of the techniques used for this purpose. Metal cations of D block elements such as rhodium, ferrous, ruthenium, copper, chromium, and nickel have been considered as metals for doping, while nonmetals such as carbon, nitrogen, fluorine, and sulfur are considered nonmetals for doping. In the doping with metals, a new donor or accepter level is inserted to decrease the bandgap, while in nonmetals doping, the shifting of the potential of VB occurs and is shifted to the upper level by inserting the new impurity level, as shown in Figure 2.5 (Rani et al. 2018).

For the effective use of the solar spectrum, the ideal bandgap of any semiconductor as a visible light photocatalyst must be nearby 2.0 eV. The VB maximum should be more positive or lower than the oxidation potential of water, while CB should be more negative than the reduction potential of hydrogen. Hence, to enhance solar photocatalysis, the tuning of the bandgap is an important criterion. Many researchers reported the shifting of absorption to the visible region by tuning the bandgap. The bandgap of TiO_2 (3.2 eV) was modified via the doping process. Likewise, other semiconductors such as $Sr_2Ta_2O_7$ (4.6 eV), $Sr_2Nb_2O_7$ (3.9 eV), $La_2Ti_2O_7$ (3.8 eV), $SrTiO_3$ (3.25 eV), $BiTaO_4$ (2.7 eV), $BiNbO_4$ (2.6 eV), WO_3 (2.6 eV), $BiVO_4$ (2.5 eV), CdS (2.42 eV), and MoS_2 (1.9 eV) have also been engineered to tune the bandgap of these for effectual visible light photocatalysis (Biswas and Baeg 2013; Gai et al. 2009; Li et al. 2013; Liu, Nisar, Ahuja, et al. 2013; Liu, Nisar, Sa, et al. 2013; Liu et al. 2012; Nisar, Pathak, et al. 2012; Nisar, Wang, et al. 2012; Wang et al. 2012).

The direct bandgaps of α-Ag_2WO_4 and α-$Ag_2W_{0.75}Mo_{0.25}O_4$ were found at 3.55 and 3.35 eV, respectively, which were calculated theoretically and were located at the Γ point of the Brillouin zone, as shown in Figure 2.6(a and b). A reduction in the bandgap values of α-$Ag_2W_{0.75}Mo_{0.25}O_4$ was observed due to the formation of a new intermediate level between VB and CB located in

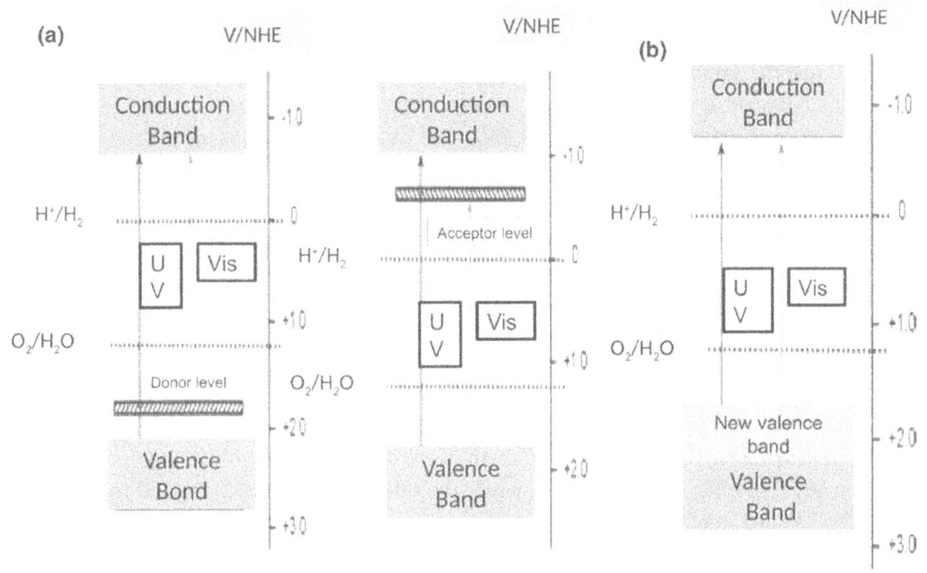

FIGURE 2.5 Schematic showing (a) the impurity level introduced in semiconductors by metal doping and (b) upshifting of valence band through nonmetal doping. (Reprinted with permission from Rani et al. (2018).)

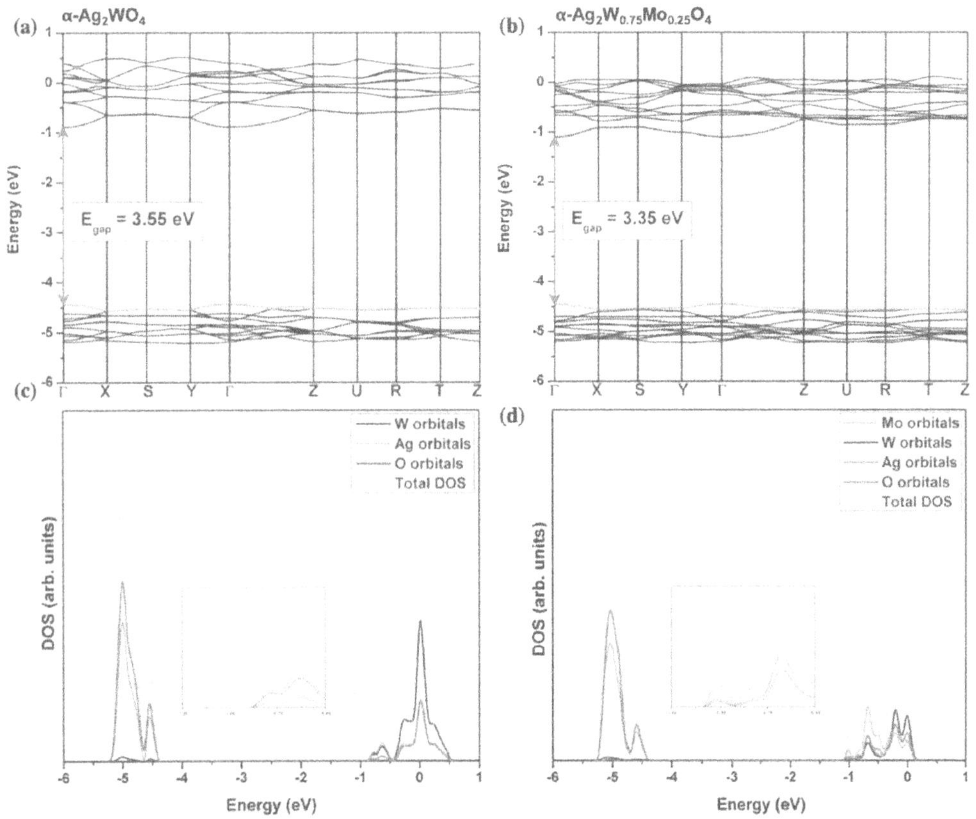

FIGURE 2.6 Band structure and density of states for the structure of (a) and (c) α-Ag_2WO_4 and (b) and (d) α-$Ag_2W_{0.75}Mo_{0.25}O_4$. (Reprinted with permission from Penha et al. (2020).)

the CB. For analyzing these new intermediate levels, the authors studied the density of states (DOS) by projecting on all the atoms, as shown in Figure 2.6(c and d). The figure depicted that the VB profile was almost the same for both but in the CB, the density intensity of α-$Ag_2W_{0.75}Mo_{0.25}O_4$ was decreased. The Mo cations in the structure of α-$Ag_2W_{0.75}Mo_{0.25}O_4$ were responsible for the decrement in bandgap in the range between −1.0 and −0.5 eV, which was absent in α-Ag_2WO_4. Hence, α-Ag_2WO_4 contributed 5d orbitals of W at the bottom of CB, while 4d orbital of Mo contributed for $Ag_2W_{0.75}Mo_{0.25}O_4$ (Penha et al. 2020).

The photodegradation of rhodamine B followed pseudo-first-order kinetics, as shown in Figure 2.7(a). The apparent velocity constant for the photodegradation was also determined by plotting the graph between $\ln(C_0/C)$ vs irradiation time and the α-$Ag_2W_{0.75}Mo_{0.25}O_4$ showed 0.023 min^{-1} high-speed constant in 16 min, which was improved as compared to pristine, as shown in Figure 2.7(b). Photodegradation of RhB was explained by the free radical tapping experiments to identify the species taking part (O_2, OH$^{\bullet}$, h$^{\bullet}$) in the degradation mechanism. The quenchers used were p-benzoquinone, $tert$-butyl alcohol, and ammonium oxalate in the degradation of the rhodamine B mechanism. The presence of p-benzoquinone suppressing the photodegradation activity suggested the key role of O_2 in the process, while the other two were not affected by this, as shown in Figure 2.7(c) (Penha et al. 2020).

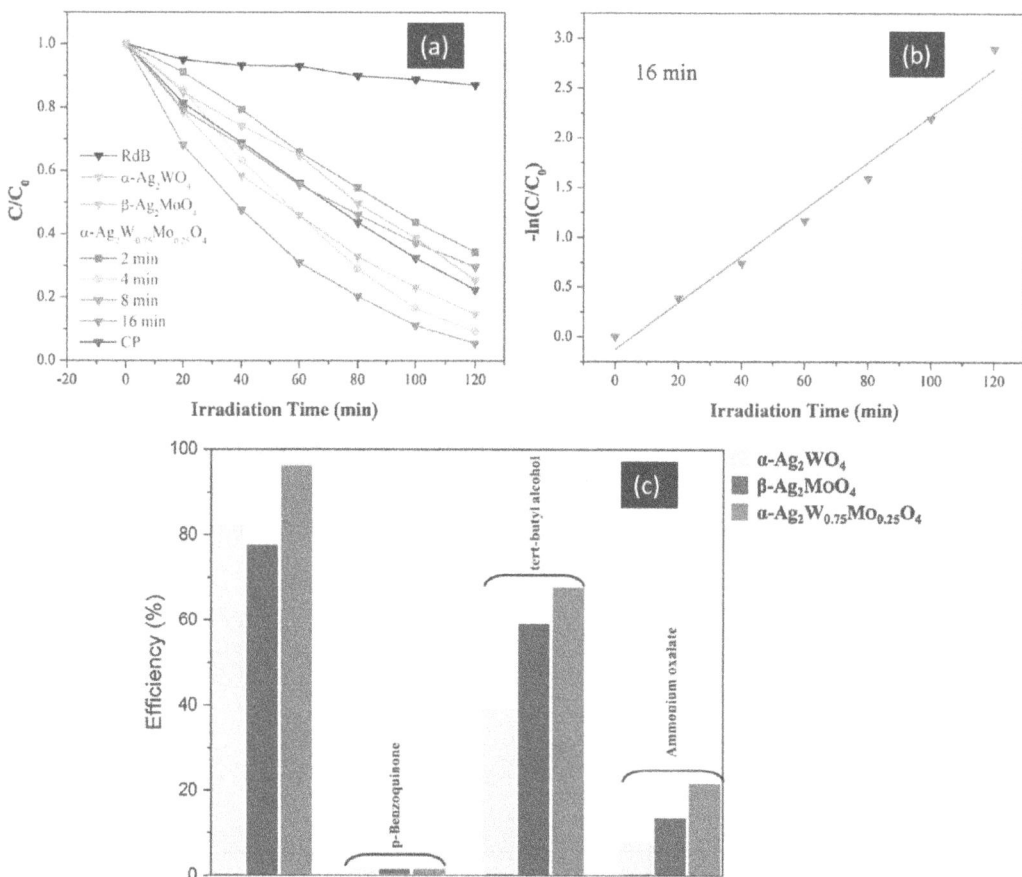

FIGURE 2.7 (a) Photocatalytic degradation of rhodamine B by α-Ag_2WO_4, β-Ag_2MoO_4, and α-$Ag_2W_{0.75}Mo_{0.25}O_4$ by microwave irradiation. (b) The kinetics studies for α-$Ag_2W_{0.75}Mo_{0.25}O_4$ in 16 min of microwave irradiation. (c) The effects of various scavengers on the visible-light photodegradation of rhodamine B for all three samples are attained at 16 min of microwave hydrothermal irradiation. (Reprinted with permission from Penha et al. (2020).)

2.7 CONCLUSION

Tuning the optical properties of semiconductors for absorption in the visible region is the key factor for solar photocatalysis. The shifting of absorption band and alignment of the bandgap are directly proportional to the photocatalytic activity. In this regard, the present chapter comprises the basic semiconductors and their hybrid materials with other inorganic-organic materials for tuning the bandgap for solar photocatalysis. The synthesis techniques for these are discussed in addition to their optical properties for enhanced photocatalysis.

ACKNOWLEDGMENTS

The author Dr Sapana Jadoun is grateful for the support of the National Research and Development Agency of Chile (ANID) and the projects, FONDECYT Postdoctoral 3200850, FONDECYT 1191572, and ANID/FONDAP/15110019. The authors are also thankful to Elsevier, Springer, American Chemical Society, Taylor & Francis, and MDPI for copyright permission.

REFERENCES

Abdel Rahman, Laila H, Ahmed M Abu-Dief, Rafat M El-Khatib, Shimaa M Abdel-Fatah, A M Adam, and E M M Ibrahim. 2018. "Sonochemical Synthesis, Structural Inspection and Semiconductor Behavior of Three New Nano Sized Cu(II), Co(II) and Ni(II) Chelates Based on Tri-dentate NOO Imine Ligand as Precursors for Metal Oxides." *Applied Organometallic Chemistry* 32 (3). Wiley Online Library: e4174.

Al-Rasheed, Radwan A. 2005. "Water Treatment by Heterogeneous Photocatalysis an Overview." In *Fourth SWCC Acquired Experience Symposium*.

Bard, Allen J. 1979. "Photoelectrochemistry and Heterogeneous Photo-Catalysis at Semiconductors." *Journal of Photochemistry* 10 (1). Elsevier: 59–75.

Bashir, Ishrat, Farooq A Lone, Rouf A Bhat, Shafat A Mir, Zubair A Dar, and Shakeel A Dar. 2020. "Concerns and Threats of Contamination on Aquatic Ecosystems." In *Bioremediation and Biotechnology*, 1–26. Springer.

Basu, Mrinmoyee, Neha Garg, and Ashok K Ganguli. 2014. "A Type-II Semiconductor (ZnO/CuS Heterostructure) for Visible Light Photocatalysis." *Journal of Materials Chemistry A* 2 (20). Royal Society of Chemistry: 7517–25.

Biswas, Soumya K, and Jin-Ook Baeg. 2013. "A Facile One-Step Synthesis of Single Crystalline Hierarchical WO_3 with Enhanced Activity for Photoelectrochemical Solar Water Oxidation." *International Journal of Hydrogen Energy* 38 (8): 3177–88. doi:https://doi.org/10.1016/j.ijhydene.2012.12.114

Chan, Maria K Y, and Gerbrand Ceder. 2010. "Efficient Band Gap Prediction for Solids." *Physical Review Letters* 105 (19). APS: 196403.

Chauvire, Timothee, Jean-Marie Mouesca, Didier Gasparutto, Jean-Luc Ravanat, Colette Lebrun, Marina Gromova, Pierre-Henri Jouneau, Jérôme Chauvin, Serge Gambarelli, and Vincent Maurel. 2015. "Redox Photocatalysis with Water-Soluble Core–Shell CdSe-ZnS Quantum Dots." *The Journal of Physical Chemistry C* 119 (31). ACS Publications: 17857–66.

Chen, Xiaobo, Yongbing Lou, Smita Dayal, Xiaofeng Qiu, Robert Krolicki, Clemens Burda, Chengfang Zhao, and James Becker. 2005. "Doped Semiconductor Nanomaterials." *Journal of Nanoscience and Nanotechnology* 5 (9). American Scientific Publishers: 1408–20.

Chiarello, Gian L, Elena Selli, and Lucio Forni. 2008. "Photocatalytic Hydrogen Production over Flame Spray Pyrolysis-Synthesised TiO_2 and Au/TiO_2." *Applied Catalysis B: Environmental* 84 (1–2). Elsevier: 332–39.

Chou, C-H, H-S Wang, K-H Wei, and Jung Y Huang. 2006. "Thiophenol-Modified CdS Nanoparticles Enhance the Luminescence of Benzoxyl Dendron-Substituted Polyfluorene Copolymers." *Advanced Functional Materials* 16 (7). Wiley Online Library: 909–16.

Colmenares, Juan C, and Y-Jun Xu. 2016. "Heterogeneous Photocatalysis." In: *Green Chemistry and Sustainable Technology*. Springer.

Cong, Chunxiao, Jingzhi Shang, Xing Wu, Bingchen Cao, Namphung Peimyoo, Caiyu Qiu, Litao Sun, and Ting Yu. 2014. "Synthesis and Optical Properties of Large-area Single-crystalline 2D Semiconductor WS2 Monolayer from Chemical Vapor Deposition." *Advanced Optical Materials* 2 (2). Wiley Online Library: 131–36.

Dong, H, Y-C Chen, and C Feldmann. 2015. "Polyol Synthesis of Nanoparticles: Status and Options Regarding Metals, Oxides, Chalcogenides, and Non-Metal Elements." *Green Chemistry* 17 (8). Royal Society of Chemistry: 4107–32.

Elilarassi, R, and G Chandrasekaran. 2012. "Synthesis and Characterization of Ball Milled Fe-Doped ZnO Diluted Magnetic Semiconductor." *Optoelectronics Letters* 8 (2). Springer: 109–12.

Elmalem, Einat, Aaron E Saunders, Ronny Costi, Asaf Salant, and Uri Banin. 2008. "Growth of Photocatalytic CdSe–Pt Nanorods and Nanonets." *Advanced Materials* 20 (22). Wiley Online Library: 4312–17.

Gai, Yanqin, Jingbo Li, Shu-Shen Li, Jian-Bai Xia, and Su-Huai Wei. 2009. "Design of Narrow-Gap TiO_2: A Passivated Codoping Approach for Enhanced Photoelectrochemical Activity." *Physical Review Letters* 102 (3). American Physical Society: 36402. doi:10.1103/PhysRevLett.102.036402

Gerion, Daniele, Fabien Pinaud, Shara C Williams, Wolfgang J Parak, Daniela Zanchet, Shimon Weiss, and A Paul Alivisatos. 2001. "Synthesis and Properties of Biocompatible Water-Soluble Silica-Coated CdSe/ZnS Semiconductor Quantum Dots." *The Journal of Physical Chemistry B* 105 (37). ACS Publications: 8861–71.

Halder, Saswata, Tushar K Bhowmik, Alo Dutta, and Tripurari Prasad Sinha. 2020. "The Photophysical Anisotropy and Electronic Structure of New Narrow Band Gap Perovskites Ln_2AlMnO_6 (Ln = La, Pr, Nd): An Experimental and DFT Perspective." *Ceramics International* 46 (13). Elsevier: 21021–32.

Hasanpoor, M, M Aliofkhazraei, and H Delavari. 2015. "Microwave-Assisted Synthesis of Zinc Oxide Nanoparticles." *Procedia Materials Science* 11. Elsevier: 320–25.

Jadoun, Sapana, Rizwan Arif, Nirmala K Jangid, and Rajesh K Meena. 2020. "Green Synthesis of Nanoparticles Using Plant Extracts: A Review." *Environmental Chemistry Letters* 19. Springer, 1–20.

Jadoun, Sapana, Anurakshee Verma, and Rizwan Arif. 2020. "Modification of Textiles via Nanomaterials and Their Applications." *Frontiers of Textile Materials*. Wiley Online Books. doi:10.1002/9781119620396.ch6

Jangid, Nirmala K, Sapana Jadoun, Anjali Yadav, Manish Srivastava, and Navjeet Kaur. 2020. "Polyaniline-TiO_2-Based Photocatalysts for Dyes Degradation." *Polymer Bulletin*. doi:10.1007/s00289-020-03318-w

Kalogirou, Soteris A. 2013. *Solar Energy Engineering: Processes and Systems*. Academic Press.

Karthikeyan, C, Prabhakarn Arunachalam, Kaliappan Ramachandran, Abdullah M Al-Mayouf, and S Karuppuchamy. 2020. "Recent Advances in Semiconductor Metal Oxides with Enhanced Methods for Solar Photocatalytic Applications." *Journal of Alloys and Compounds* 828. Elsevier: 154281.

Kubacka, Anna, Marcos Fernandez-Garcia, and Gerardo Colon. 2012. "Advanced Nanoarchitectures for Solar Photocatalytic Applications." *Chemical Reviews* 112 (3). ACS Publications: 1555–1614.

Li, Yunguo, Yan-Ling Li, Carlos M Araujo, Wei Luo, and Rajeev Ahuja. 2013. "Single-Layer MoS_2 as an Efficient Photocatalyst." *Catalysis Science & Technology* 3 (9). The Royal Society of Chemistry: 2214–20. doi:10.1039/C3CY00207A

Li, Zizhen, Xiangchao Meng, and Zisheng Zhang. 2018. "Recent Development on MoS_2-Based Photocatalysis: A Review." *Journal of Photochemistry and Photobiology C: Photochemistry Reviews* 35: 39–55. doi:10.1016/j.jphotochemrev.2017.12.002

Liao, Guangfu, Yan Gong, Li Zhang, Haiyang Gao, Guan-Jun Yang, and Baizeng Fang. 2019. "Semiconductor Polymeric Graphitic Carbon Nitride Photocatalysts: The 'Holy Grail' for the Photocatalytic Hydrogen Evolution Reaction under Visible Light." *Energy & Environmental Science* 12 (7). Royal Society of Chemistry: 2080–2147.

Lincot, Daniel. 2005. "Electrodeposition of Semiconductors." *Thin Solid Films* 487 (1–2). Elsevier: 40–48.

Liu, Chun-Sheng, Xiao-Le Yang, Jin Liu, and Xiao-Juan Ye. 2018. "Exfoliated Monolayer GeI2: Theoretical Prediction of a Wide-Band Gap Semiconductor with Tunable Half-Metallic Ferromagnetism." *The Journal of Physical Chemistry C* 122 (38). ACS Publications: 22137–42.

Liu, Peng, Jawad Nisar, Rajeev Ahuja, and Biswarup Pathak. 2013. "Layered Perovskite $Sr_2Ta_2O_7$ for Visible Light Photocatalysis: A First Principles Study." *The Journal of Physical Chemistry C* 117 (10). American Chemical Society: 5043–50. doi:10.1021/jp310945e

Liu, Peng, Jawad Nisar, Biswarup Pathak, and Rajeev Ahuja. 2012. "Hybrid Density Functional Study on $SrTiO_3$ for Visible Light Photocatalysis." *International Journal of Hydrogen Energy* 37 (16): 11611–17. doi:10.1016/j.ijhydene.2012.05.038

Liu, Peng, Jawad Nisar, Baisheng Sa, Biswarup Pathak, and Rajeev Ahuja. 2013. "Anion–Anion Mediated Coupling in Layered Perovskite $La_2Ti_2O_7$ for Visible Light Photocatalysis." *The Journal of Physical Chemistry C* 117 (27). American Chemical Society: 1384552. doi:10.1021/jp402971b

Liu, Xin, and Mark T Swihart. 2014. "Heavily-Doped Colloidal Semiconductor and Metal Oxide Nanocrystals: An Emerging New Class of Plasmonic Nanomaterials." *Chemical Society Reviews* 43 (11). Royal Society of Chemistry: 3908–20.

Liu, Yanping, Shijie Shen, Jitang Zhang, Wenwu Zhong, and Xiaohua Huang. 2019. "$Cu_{2-x}Se/CdS$ Composite Photocatalyst with Enhanced Visible Light Photocatalysis Activity." *Applied Surface Science* 478. Elsevier: 762–69.

Makuła, Patrycja, Michał Pacia, and Wojciech Macyk. 2018. "How to Correctly Determine the Band Gap Energy of Modified Semiconductor Photocatalysts Based on UV–Vis Spectra." ACS Publications.

Marschall, Roland. 2014. "Semiconductor Composites: Strategies for Enhancing Charge Carrier Separation to Improve Photocatalytic Activity." *Advanced Functional Materials* 24 (17). John Wiley & Sons, Ltd: 2421–40. doi:10.1002/adfm.201303214

Nakata, Kazuya, and Akira Fujishima. 2012. "TiO_2 Photocatalysis: Design and Applications." *Journal of Photochemistry and Photobiology C: Photochemistry Reviews* 13 (3): 169–89. doi:10.1016/j.jphotochemrev.2012.06.001

Nisar, Jawad, Biswarup Pathak, Baochang Wang, Tae W Kang, and Rajeev Ahuja. 2012. "Hole Mediated Coupling in $Sr_2Nb_2O_7$ for Visible Light Photocatalysis." *Physical Chemistry Chemical Physics* 14 (14). The Royal Society of Chemistry: 4891–97. doi:10.1039/C2CP23912D

Nisar, Jawad, Baochang Wang, Carlos M Araujo, Antonio F da Silva, Tae W Kang, and Rajeev Ahuja. 2012. "Band Gap Engineering by Anion Doping in the Photocatalyst $BiTaO_4$: First Principle Calculations." *International Journal of Hydrogen Energy* 37 (4): 3014–18. doi:10.1016/j.ijhydene.2011.11.068

Pelaez, Miguel, Nicholas T Nolan, Suresh C Pillai, Michael K Seery, Polycarpos Falaras, Athanassios G Kontos, Patrick S M Dunlop, Jeremy W J Hamilton, John Anthony Byrne, and Kevin O'shea. 2012. "A Review on the Visible Light Active Titanium Dioxide Photocatalysts for Environmental Applications." *Applied Catalysis B: Environmental* 125. Elsevier: 331–49.

Penha, M D, A F Gouveia, M M Teixeira, R C de Oliveira, M Assis, J R Sambrano, F Yokaichya, et al. 2020. "Structure, Optical Properties, and Photocatalytic Activity of α-$Ag_2W_{0.75}Mo_{0.25}O_4$." *Materials Research Bulletin* 132: 111011. doi:10.1016/j.materresbull.2020.111011

Rani, Ankita, Rajesh Reddy, Uttkarshni Sharma, Priya Mukherjee, Priyanka Mishra, Aneek Kuila, Lan C Sim, and Pichiah Saravanan. 2018. "A Review on the Progress of Nanostructure Materials for Energy Harnessing and Environmental Remediation." *Journal of Nanostructure in Chemistry* 8 (3). Springer: 255–91.

Rashid, Haroon Ur, Kaichao Yu, Muhammad N Umar, Muhammad N Anjum, Khalid Khan, Nasir Ahmad, and Muhammad T Jan. 2015. "Catalyst Role in Chemical Vapor Deposition (CVD) Process: A Review." *Reviews on Advanced Materials Science* 40 (3): 235–48.

Robert, Didier, and Sixto Malato. 2002. "Solar Photocatalysis: A Clean Process for Water Detoxification." *Science of the Total Environment* 291 (1–3). Elsevier: 85–97.

Schneider, Jenny, Masaya Matsuoka, Masato Takeuchi, Jinlong Zhang, Yu Horiuchi, Masakazu Anpo, and Detlef W Bahnemann. 2014. "Understanding TiO_2 Photocatalysis: Mechanisms and Materials." *Chemical Reviews* 114 (19). ACS Publications: 9919–86.

Seeger, Torsten, Philipp Kohler-Redlich, and Manfred Ruehle. 2000. "Synthesis of Nanometer-Sized SiC Whiskers in the Arc-Discharge." *Advanced Materials* 12 (4). Wiley Online Library: 279–82.

Shen, Shaohua, Jinwen Shi, Penghui Guo, and Liejin Guo. 2011. "Visible-Light-Driven Photocatalytic Water Splitting on Nanostructured Semiconducting Materials." *International Journal of Nanotechnology* 8 (6–7). Inderscience Publishers: 523–91.

Shi, Linlin, Keqiang Chen, Aiping Zhai, Guohui Li, Mingming Fan, Yuying Hao, Furong Zhu, Han Zhang, and Yanxia Cui. 2021. "Status and Outlook of Metal–Inorganic Semiconductor–Metal Photodetectors." *Laser & Photonics Reviews* 15 (1). Wiley Online Library: 2000401.

Spasiano, Danilo, Raffaele Marotta, Sixto Malato, Pilar Fernandez-Ibañez, and Ilaria Di Somma. 2015. "Solar Photocatalysis: Materials, Reactors, Some Commercial, and Pre-Industrialized Applications. A Comprehensive Approach." *Applied Catalysis B: Environmental* 170–171: 90–123. doi:10.1016/j.apcatb.2014.12.050

Suresh, P, P Balaraju, S K Sharma, M S Roy, and G D Sharma. 2008. "Photovoltaic Devices Based on PPHT: ZnO and Dye-Sensitized PPHT: ZnO Thin Films." *Solar Energy Materials and Solar Cells* 92 (8). Elsevier: 900–908.

Tang, K-B, Y-T Qian, J-H Zeng, and X-G Yang. 2003. "Solvothermal Route to Semiconductor Nanowires." *Advanced Materials* 15 (5). Wiley Online Library: 448–50.

Tolia, Jyoti, Mousumi Chakraborty, and Z V P Murthy. 2012. "Synthesis and Characterization of Semiconductor Metal Sulfide Nanocrystals Using Microemulsion Technique." *Crystal Research and Technology* 47 (8). Wiley Online Library: 909–16.

Tsay, Chien-Yie, Chun-Wei Wu, Chien-Ming Lei, Fan-Shiong Chen, and Chung-Kwei Lin. 2010. "Microstructural and Optical Properties of Ga-Doped ZnO Semiconductor Thin Films Prepared by Sol–Gel Process." *Thin Solid Films* 519 (5). Elsevier: 1516–20.

Usman, Muhammad, Shruti Mendiratta, and Kuang-Lieh Lu. 2017. "Semiconductor Metal–Organic Frameworks: Future Low-Bandgap Materials." *Advanced Materials* 29 (6). Wiley Online Library: 1605071.

Verma, Anurakshee, Rizwan Arif, and Sapana Jadoun. 2020. "Synthesis, Characterization, and Application of Modified Textile Nanomaterials." In: Frontiers of Textile Materials. Wiley Online Books. doi:10.1002/9781119620396.ch8

Wang, B C, J Nisar, B Pathak, T W Kang, and R Ahuja. 2012. "Band Gap Engineering in BiNbO4 for Visible-Light Photocatalysis." *Applied Physics Letters* 100 (18). American Institute of Physics: 182102. doi:10.1063/1.4709488

Zhou, Chengyun, Cui Lai, Chen Zhang, Guangming Zeng, Danlian Huang, Min Cheng, Liang Hu, Weiping Xiong, Ming Chen, and Jiajia Wang. 2018. "Semiconductor/Boron Nitride Composites: Synthesis, Properties, and Photocatalysis Applications." *Applied Catalysis B: Environmental* 238. Elsevier: 6–18.

Zou, Guifu, Hui Li, Yuanguang Zhang, Kan Xiong, and Yitai Qian. 2006. "Solvothermal/Hydrothermal Route to Semiconductor Nanowires." *Nanotechnology* 17 (11). IOP Publishing: S313.

3 Semiconductor Optical Memory Devices

Umbreen Rasheed, Fayyaz Hussain,
Rana Muhammad Arif Khalil, and Muhammad Imran

CONTENTS

3.1 INTRODUCTION

Optical memory devices are of increasing interests due to their potential applications in information processing, memory computing, data transmitting and image capturing applications [1]. This interest is boosted up due to high-speed signal processing application of light in optical fibers [2]. Initial attempts give rise to flip-flop optical memory devices based on pocket level processing. Progressing levels lead to the expansion of functionality and applications of optical memory devices based on bit-level storage during the last decades. Bit-level storage devices are further categorized as volatile and non-volatile depending on their data storage capability on turning off the power. High-speed operations and faster time access of volatile optical memories brought them closer to control processing unit (CPU). Speed and energy advantages of light penetrated the use of optical random-access memories (RAMs) in computers [3]. More appealing features of these devices are interchangeability, affordability, removability, reusability, durability and data stability. Compact disc (CD) being a read-only memory (ROM) device, digital versatile disc (DVD) ROM, Blu-ray high-density DVD and DVD random-access memory (RAM) are examples of optical memory devices. The use of old optical memory devices (CD-R and DVD-R) was limited due to their capability of burning data just once, that is, user was unable to reuse it by deleting or erasing the stored data. Nowadays, rewrite and rereadable optical memory devices (CD-RW and DVD-RW) are also available in market [4].

Neutral nature of photons limits the practical application of optical memories in electronic-capacitor-based resistive RAM (RRAM) devices. The most common approach is to rely on bistable optical resonance (artificial cavities) or bistable nature of device (material-based approach). Optical memory devices store data in the form of dots and these data can be read using light or specially laser light. In such devices, all read and write activities are performed by light. Light reflected from these dots behave differently and data are clearly read. It is interesting to mention that these dots are created by laser light in high-power mode marking on the surface of material [5]. Basic principle of optical memory devices is based on charge trapping and de-trapping in defect sites in the material used.

Memory operation is fulfilled with the achievement of two stable states, i.e., logical zero and one. Zero is more commonly defined as a high resistance state (HRS), present in the region of no emission of light, whereas one is a low resistance state (LRS), present in the region of light emission [6]. Completion of this operation is done by allowing the switching between these two states.

DOI: 10.1201/9781003188582-3

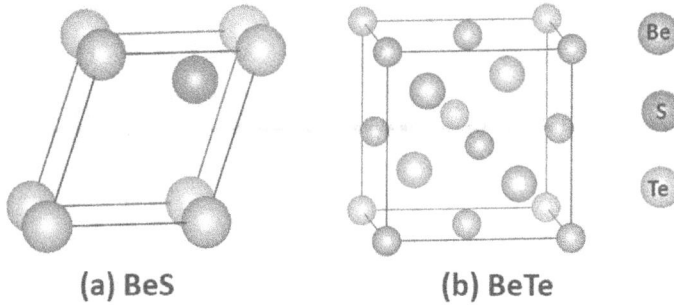

(a) BeS   **(b) BeTe**

FIGURE 3.1 Optimized structures of II–VI semiconductors (a) BeS and (b) BeTe.

Photo-switchable materials are largely used in optical memory devices due to their image sensing, photo processing and optical recording capability [7–9]. Semiconductors have been important candidates as optical memory materials due to their existence in bistable states. In the absence of any external stimuli, with a certain value of bandgap, HRS (logical zero) is attained. LRS (logical one) may be attained with photogenerated electrons with light stimuli. Photoresponsivity of semiconductors due to the possibility of photogenerated electrons in the presence of light stimuli has made them more appealing in the practical applications of optical memory devices [10].

Nowadays, unique properties of semiconductors (graphene, transition metal oxides, transition metal dichalcogenides, II–VI and III–V) are making these upcoming building blocks of new generation electronic and optoelectronic devices [6]. Kim *et al.* reported the important applications of semiconductors like MoS_2 and graphene as floating gates to be used in optical memory devices [1]. Semiconductors due to their bandgap may exist in bistable states. This feature makes them suitable for optical memory devices. Graphene is another interesting semiconductor with its important applications in optical modulators and data processing devices [6, 11].

3.2 II–VI AND III–V SEMICONDUCTORS

The use of II–VI and III–V group semiconductors in optoelectronic devices is increasing day by day due to the improvement in efficiency. BeS and BeTe shown in Figure 3.1(a and b) are two important composites belonging to II–VI group of semiconductors, whereas BN and GaAs shown in Figure 3.2(a and b) are the composites belonging to III–V group semiconductors. Here, properties of

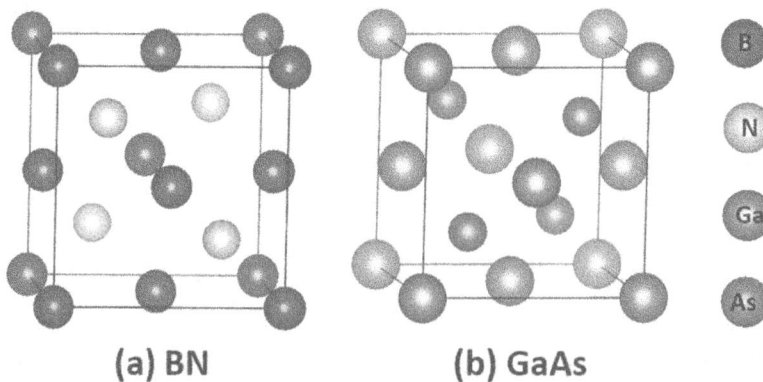

(a) BN   **(b) GaAs**

FIGURE 3.2 Optimized structures of III–V semiconductors (a) BN and (b) GaAs.

TABLE 3.1

Parameters Used for Modeling II–VI and III–V Semiconductors

Parameters Used	II–VI Semiconductor		III–V Semiconductor	
	BeS	BeTe	BN	GaAs
Plane wave basis set cutoff (eV)	180	326	220	299.3000
MP grid size for SCF calculations	4 4 4	4 4 4	6 6 6	4 4 4
Point group	30: Td, $\bar{4}$3m, $\bar{4}$ 3m	6: C2v, mm2, 2 mm	30: Td, $\bar{4}$3m, $\bar{4}$ 3m	30: Td, $\bar{4}$3m, $\bar{4}$ 3m
Space group	216: F$\bar{4}$3m, F $\bar{4}$ 2 3	44: Imm2, I 2 -2	216: F$\bar{4}$3m, F $\bar{4}$ 2 3	216: F$\bar{4}$3m, F $\bar{4}$ 2 3

these four composites are investigated using density functional theory (DFT) employing Vienna Ab Initio Simulation Package (VASP) code. Calculated results like energy bandgap (Eg) and refractive index may be underestimated due to under approximation of DFT. These results may assist in better understanding of significance of semiconductors for optical memory applications.

The structural and electronic properties of the optimized structures have been determined using Perdew, Burke and Ernzerhof (PBE) functionals under generalized gradient approximation (GGA) [12–14]. The valence states of Be $(2s^2)$, B $(2s^2\ 2p^1)$, Ga $(3d^{10}\ 4s^2\ 4p^1)$, S $(3s^2\ 3p^4)$, Te $(5s^2\ 5p4)$, N $(2s^2\ 2p^3)$ and As $(4s^2\ 4p^3)$ were considered for calculations. All these calculations have been carried out employing VASP [12, 13]. For converging, the total energy of the materials w.r.t. electron wave functions, plane waves with cut-off energy listed in Table 3.1 have been utilized. The lattice parameters, cell volume and atomic sites have been optimized after complete relaxation through conjugate gradient (CG) approximation so that Hellmann-Feynman forces turned out to be < 0.02 eV/Å, and to meet ~1 × 10^{-5} eV for energy convergence [15]. The Monkhorst-Pack (MP) [16, 17] was used for sampling k-points. The MP grid size given in Table 3.1 has been selected in all investigations. All calculated equilibrium parameters and results of mechanical properties of the four optimized composites are tabulated in Table 3.2.

3.2.1 ELECTRONIC PROPERTIES

Electronic properties play a crucial role in determining the nature of the material and hence predicting its specific use for a particular application. These properties assist to analyze the basic requirement of optical memory devices, i.e., their existence in bistable state. In theoretical physics,

TABLE 3.2

Structural and Mechanical Parameters of Optimized Structures

Measured Parameter	II–VI Semiconductor		III–V Semiconductor	
	BeS	BeTe	BN	GaAs
Compressibility (1/GPa)	0.0116	0.0237	0.0038	0.0021
Young's modulus (GPa) along X = Y = Z	108.4736	32.8728	672.3165	290.1477
Bulk modulus (GPa)	86.4528	42.1314	266.5754	474.0139
Shear modulus (Lame Mu)	370.1610	40.7217	370.1610	−562.6230
Poisson ratios, Exy = Eyz = Ezx	0.2909	0.3700	0.0797	0.3980
Universal anisotropy index:	0.7927	3.8287	0.0904	−14.1669
Lattice parameters (Å) a = b = c	3.4401	3.9782	2.5562	3.9974
Current cell volume (Å^3)	28.7865	44.5184	11.8104	45.1660

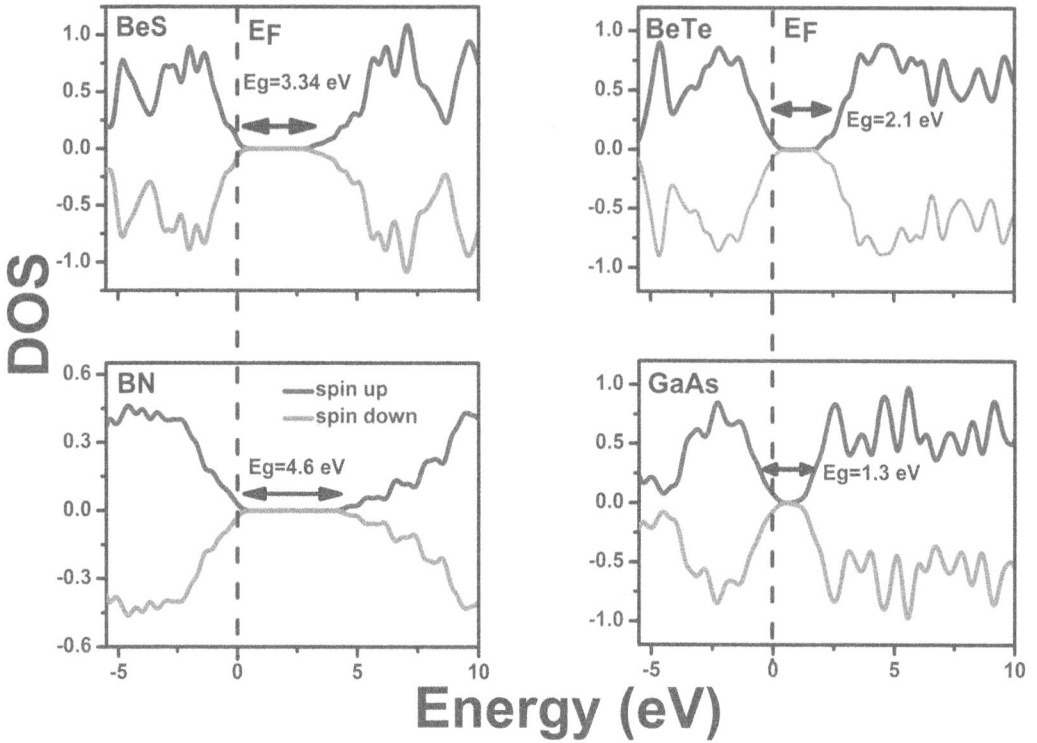

FIGURE 3.3 Density of states plots of the four composites. Vertical, dashed black line indicting Fermi level.

density of states (DOS) versus energy plots are used for the description of electronic properties. DOS plots of the four composites are shown in Figure 3.3. In all DOS plots, dashed line is indicating Fermi level for the highest occupied energy level, whereas left and right side of this Fermi level is illustrating valence band and conduction band, respectively. Bandgaps of the four composites calculated using VASP are summarized in Table 3.3. In DOS plots of the four composites, bandgap with no defect states indicates the insulating nature. DOS plots in Figure 3.3 for II–VI (BeS, BeTe) and III–V (BN, GaAs) semiconductors indicate their existence in HRS. Measured bandgap also predicts the energy of incident light radiations required to switch the optical memory devices based on these materials to LRS. Incident light radiation being an external stimulus may create photogenerated electrons switching the device into conducting state. In this way, these four

TABLE 3.3

Energy Bandgap, Static Dielectric Constant and Static Refractive Index of the Optimized Structures

Parameters Measured	II–VI Semiconductor		III–V Semiconductor	
	BeS	BeTe	BN	GaAs
Energy bandgap (eV)	3.34	2.1	4.6	1.3
Static dielectric constant	6.9	8.6	4.5	14.8
Static refractive index	2.3	2.9	2.1	3.9

composites may be used to achieve bistable states in optical memory devices. This bandgap may be treated as potential well. Higher bandgap may be used as deeper potential well having capability to store a greater number of electrons. Higher bandgap of BN is depicting its most suitable candidate for optical memory applications. This is due to the capability to create more charge trapping centers with longer lifetime. Longer lifetime is attributed to their generation deep inside the energy gap when exposed to light.

3.2.2 OPTICAL PROPERTIES

Optical properties of the material play much crucial role in predicting its optoelectronic applications. Absorption of light plays an important role in deciding the existence of device in specific logical state. Here, optical properties of the four composites interacting with electromagnetic radiations were investigated within the incident energy range of 0–40 eV. Absorption is an important optical property that results in an increased population of excited states [9]. As polarized light in specific direction is allowing the light to emit in a particular direction, semiconductors show this unique property depending on the value and nature (direct or indirect) of their bandgap. Here, results of absorbed light along [100] the direction are shown in Figure 3.4. These indicate that semiconducting materials simulated from II–VI group (BeS, BeTe) and III–V group (BN, GaAs) have zero absorbance in low-energy infrared region and high-energy ultraviolet region. Zero absorption in infrared region is confirming the existence of bandgap shown in DOS plots. This will help in achieving the HRS in optical memory devices with such materials in the absence of light stimuli. However, absorption peaks are showing the possibility of photoresponsivity creating photogenerated electrons. This will assist in achieving the other bistable state of LRS in such

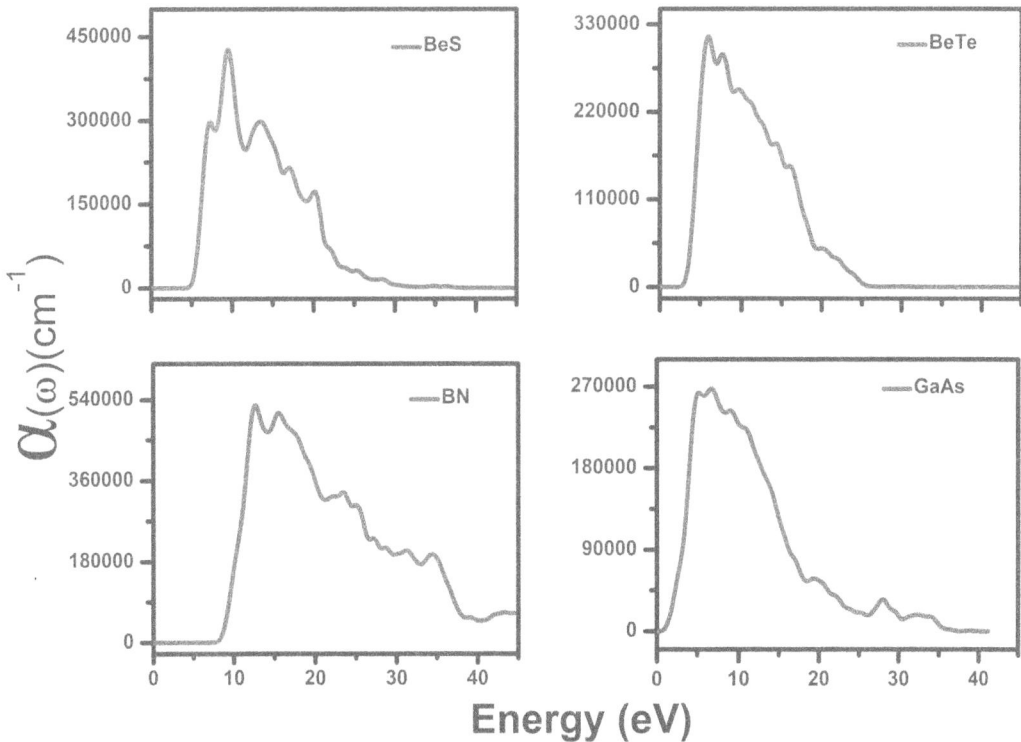

FIGURE 3.4 Absorption coefficient versus energy plots of the four composites.

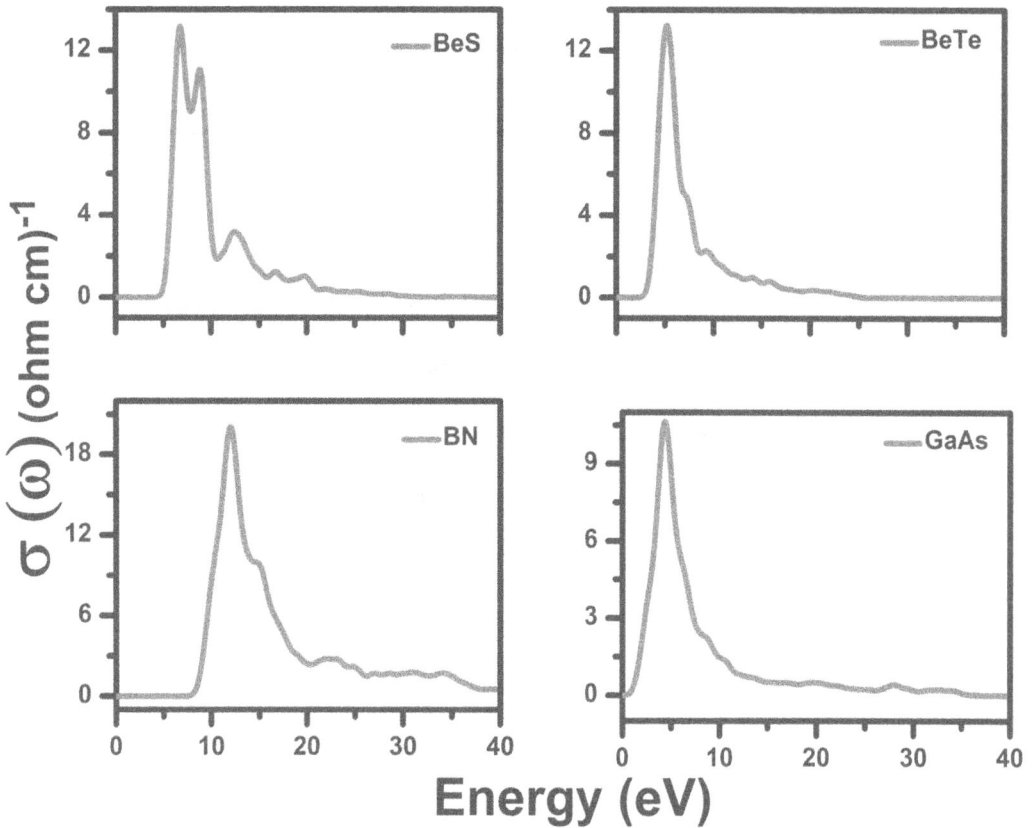

FIGURE 3.5 Optical conductivity spectra of the four composites.

materials in the presence of light stimuli. This absorption phenomenon may assist in predicting the image sensing capability of the four composites while exposed to light. Image sensing capability may be identified by the absorption spectrum, which identifies the range of energy. The absorption peaks will identify the exposed dose needed for image sensing. Maximum absorption peak showed comparatively higher value of image sensing when composite was exposed to light. This factor may also be used for writing due to the creation of more excited states. Results in Figure 3.4 are showing that image sensing capability of BN is comparatively higher than the other three composites due to its higher absorption peak.

To check the photoresponsivity of the four composites, effects of incidental electromagnetic radiations within infrared, visible and low-energy ultraviolet region on optical conductivity are investigated. Optical conductivity response of the four composites is shown in Figure 3.5. Zero conductivity in low-energy region, i.e., infrared radiations as external stimuli, shows the existence of optical memory devices based on the BeS, BeTe, BN and GaAs materials in HRS. This confirms the existence of bandgap existing in these semiconductors. Zero conductivity range extending within lesser energy range (in infrared region) confirms the GaAs as a semiconducting material with lesser bandgap as shown in DOS plots, whereas BN is shown with comparatively higher bandgap. Optical conductivity within visible light region confirms the sensitivity of the four semiconducting composites for light. This optical conductivity is due to photogenerated electrons in the presence of light as external stimuli. This conductivity will lead these composites into LRS. In this way, bistable states will be achieved in the optical memory devices based on these materials. Lee *et al.* reported the photogenerated electron-hole pairs in MoS_2 monolayer when exposed to laser light of 450 nm [18].

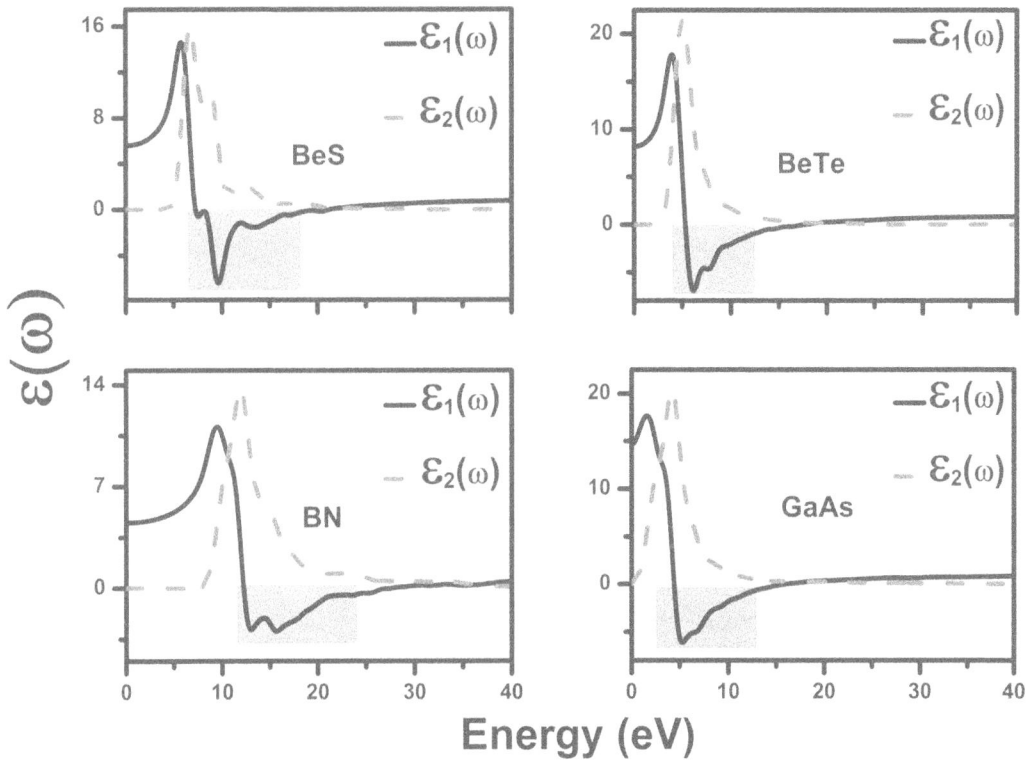

FIGURE 3.6 Dielectric constant versus energy plots of the four composites. Black solid and red dashed lines indicate real and imaginary parts of dielectric constant. Highlighted rectangular portion indicates negative values of the imaginary part of dielectric constant.

They also reported that with an increasing number of photogenerated electrons, more electrons will be transferred to the conduction band.

Dielectric constant (ε) is the central optical property deciding the practical optoelectronic applications of the composite. Real and imaginary dielectric constants (ε_1 and ε_2) versus energy plots of the four composites are displayed in Figure 3.6. Static values of (ε_1) at 0 eV are tabulated in Table 3.3. It is clear that all composites have metallicity due to static positive values. It is clear that ε_1 increases as the energy of incident electromagnetic radiations increases. Negative value of ε_1 confirms the insulating nature displayed by these composites.

Refractive index (summation of real and imaginary parts of refractive index) is an important parameter while selecting material for designing optical memory device. Change in refractive index (n) plays a crucial role in the presence of light stimulus [5]. Static values of refractive index (n at 0 eV) are tabulated in Table 3.3. These results show that refractive index of GaAs is comparatively higher than the other composites, whereas that of BN is lower than that of other studied composites. To investigate the influence of incident electromagnetic radiations on n, refractive index versus incident energy plots of these four composites are shown in Figure 3.7. An increase in refractive index with external stimulus shifts the phase paradigm from crystalline to amorphous and vice versa. Results of the four semiconductors in Figure 3.7 demonstrate this phase change due to increasing refractive via interaction with light in visible region. These results also confirm the suitability of these four semiconducting composites for phase-change memory (PCM) devices. In Figure 3.7, solid black and dashed blue lines are indicative of real (n_1) and imaginary (n_2) parts of refractive index, respectively.

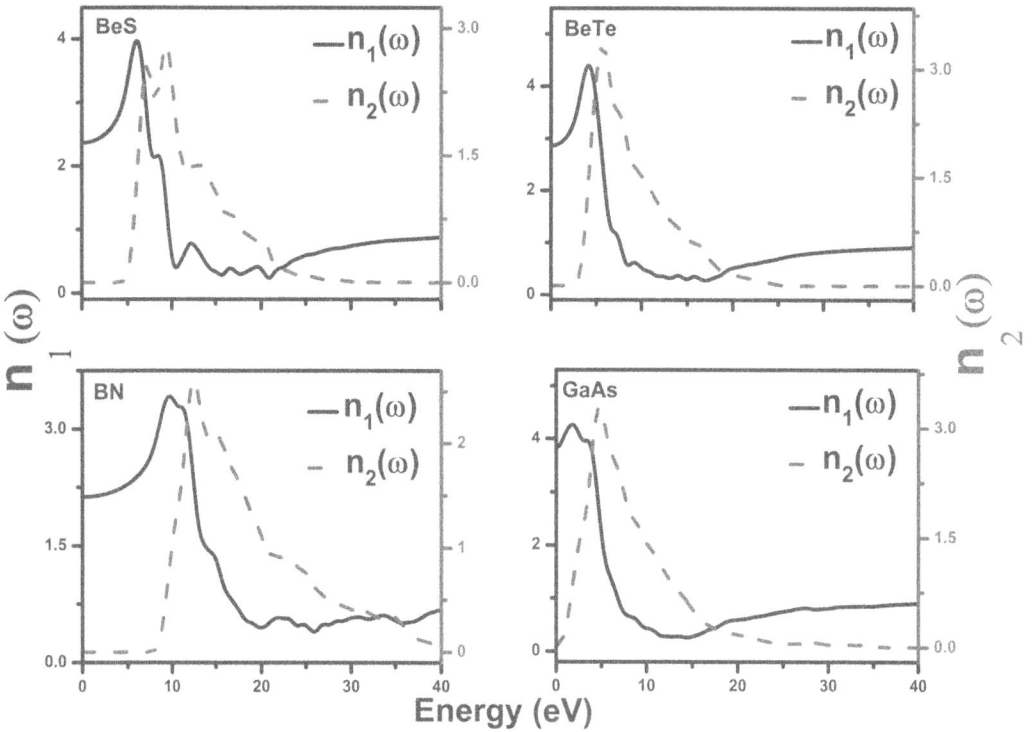

FIGURE 3.7 Refractive index versus energy plots of the four composites. Black solid and blue dashed lines indicate real and imaginary parts of refractive index.

3.3 ZnO

ZnO is a wide bandgap semiconductor. This ability makes this composite suitable for devices having ability to sustain high electric fields and operate at high power and temperature. Cubic structure of this semiconductor is shown in Figure 3.8(a). Its theoretically calculated bandgap shown in Figure 3.8(b) is slightly lesser than experimentally reported bandgap due to the under-approximation of DFT. ZnO in its pure form will be an insulator at room temperature. This state is identified as HRS in optical memory devices.

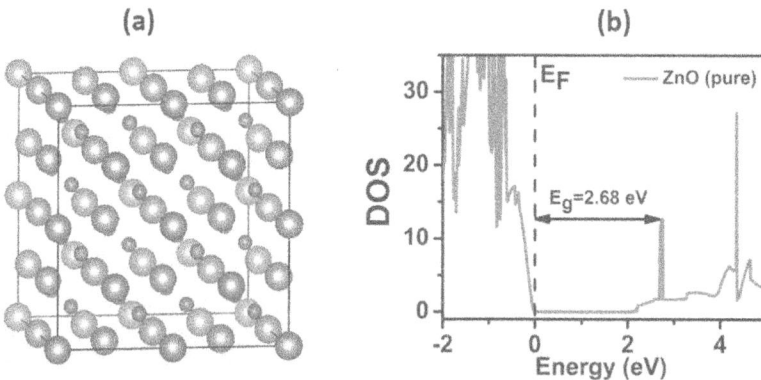

FIGURE 3.8 ZnO (a) optimized structure; gray and red color balls representing zinc and oxygen atoms and (b) DOS versus energy plot.

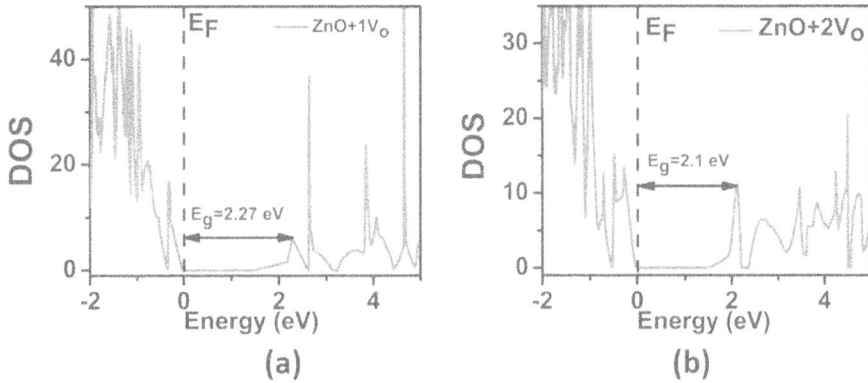

FIGURE 3.9 DOS versus energy plot for (a) ZnO + $1V_o$ and (b) ZnO + $2V_o$.

To investigate its applicability for optical memory devices, effect of oxygen vacancy (V_o) as a charge trapping center was examined. For this purpose, single V_o ($1V_o$) and di-V_o ($2V_o$) were modeled in cubic ZnO composites. Ke *et al.* [19] also demonstrated the effect of intrinsic and extrinsic defects in the material for optical memory devices. It is clear from DOS of ZnO + $1V_o$ in Figure 3.9(a) that bandgap of ZnO decreases in the presence of $1V_o$. The number of localized states in conduction band also increased due to $1V_o$. In this way, conductivity of ZnO increased in the presence of $1V_o$ that switches it into LRS. This effect increased further by increasing the number of V_os ($2V_o$ in this case) as shown in Figure 3.9(b). This shows that a number of unpaired electron-hole pairs are increased due to V_os. This created accumulation of electron-hole pairs in conduction band. This reflects the suitability of ZnO semiconductor-based optical memory devices. In the presence of light, optically generated electrons may be trapped in these trapping centers. More trapping centers with increasing V_os may store a greater number of electrons in potential created in conduction band for longer lifetime. After waiting for some time in darkness, readout biasing applied to such optical memory devices will restore all charges and bring ZnO composite to its original HRS.

REFERENCES

1. Sung H. Kim, Sum-Gyun Yi, Myung U. Park, ChangJun Lee, Myeongjin Kim, and Kyung-Hwa Yoo. 2019. "Multilevel MoS$_2$ optical memory with photoresponsive top floating gates." *ACS Appl. Mater. Interfaces* 11: 25306–25312.
2. Martin T. Hill, Harmen J. S. Dorren, Tjibbe de Vries, Xaveer J. M. Leijtens, Jan H. den Besten, Barry Smalbrugge, Yok-Siang Oei, Hans Binsma, Giok-Djan Khoe, and Meint K. Smit. 2004. "A fast low-power optical memory based on coupled micro-ring lasers." *Nature* 432: 206–209.
3. Theoni Alexoudi, George T. Kanellos, and Nikos Pleros. 2020. "Optical RAM and integrated optical memories: a survey." *Light: Sci. Appl.* 9: 91.
4. Yoshinobu Mitsuhashi. 1998. "Optical storage: Science and technology." *Jpn. J. Appl. Phys.* 37: 2079–2083.
5. Fuxi Gan, Lisong Hou, Guangbin Wang, Huiyong Liu, and Jing Li. 2000. "Optical and recording properties of short wavelength optical storage materials." *Mater. Sci. Eng. B.* B76: 63–68.
6. Che-Wei Chang, Wei-Chun Tan, Meng-Lin Lu, Tai-Chun Pan, Ying-Jay Yang, and Yang-Fang Chen. 2014. "Electrically and optically readable light emitting memories." *Sci. Rep.* 4: 512
7. Y. Zhou, K. S. Yew, D. S. Ang, T. Kawashima, M. K. Bera, H. Z. Zhang, and G. Bersuker. 2015. "White-light-induced disruption of nanoscale conducting filament in hafnia." *Appl. Phys. Lett.* 107, 072107.
8. Jacky C.-H. Chan, Wai H. Lam, and Vivian W.-W. Yam. 2014. "A highly efficient silole-containing dithienylethene with excellent thermal stability and fatigue resistance: A promising candidate for optical memory storage materials." *J. Am. Chem. Soc.* 136: 16994–16997.
9. A. S. Dvornikov, J. Makin, and P. M. Rentzepis. 1994. "Spectroscopy and kinetics of photochromic materials for 3D optical memory devices." *J. Phys. Chem.* 98: 61466752.

10. Noboru Yamada. 1996. "Erasable phase-change optical materials." *MRS Bull.* 21: 48–50.

11. Kallol Roy, Medini Padmanabhan, Srijit Goswami, T. P. Sai, Gopalakrishnan Ramalingam, Srinivasan Raghavan, and Arindam Ghosh. 2013. "Graphene–MoS_2 hybrid structures for multifunctional photoresponsive memory devices." *Nat. Nanotechnol.* 8: 826–830.

12. J. P. Perdew, K. Burke, and M. Ernzerhof. 1997. "Generalized gradient approximation made simple." *Phys. Rev. Lett.* 77: 3865.

13. P. E. Blochl. 1993. "Projector augmented-wave method." *Phys. Rev. B.* 47: 558.

14. G. Kress, and D. Joubert. 1999. "From ultrasoft pseudopotentials to the projector augmented-wave method." *Phys. Rev. B.* 59: 1758–1775.

15. A. V. Krukau, O. A. Vydrov, A. F. Izmaylov, and G. E. Scuseria. 2006. "Influence of the exchange screening parameter on the performance of screened hybrid functional." *J. Chem. Phys.* 125: 224106.

16. H. J Monkhorst, and J. D. Pack. 1976. "Special points for Brillouin-zone integrations." *Phys Rev B.* 13: 5188–5192.

17. H. J. Monkhorst, and J. D. Pack. 1977. "Special points for Brillouin-zone integrations: a reply." *Phys Rev B.* 16: 1748.

18. Juwon Lee, Sangyeon Pak, Young-Woo Lee, Yuljae Cho, John Hong, Paul Giraud, Hyeon S. Shin, Stephen M. Morris, Jung I. Sohn, Seung Nam Cha, and Jong M. Kim. 2017. "Monolayer optical memory cells based on artificial trap-mediated charge storage and release." *Nat. Commun.* 8: 14734.

19. Ke Pei, Xiaochen Ren, Zhiwen Zhou, Zhichao Zhang, Xudong Ji, and Paddy K. L. Chan. 2018. "A high-performance optical memory array based on inhomogeneity of organic semiconductors." *Adv. Mater.* 30: 1706647.

4 Semiconductor Optical Utilization in Agriculture

Syed Wazed Ali, Satyaranjan Bairagi, and Swagata Banerjee

CONTENTS

4.1 INTRODUCTION

Agriculture has been one of the major sectors experiencing development in recent years. With the population increasing exponentially in many countries, the difficulties in meeting the agricultural demand are also rising. To keep at par with the demanding situation, various types of chemicals in the form of pesticides and fertilizers are being employed. Some of these fertilizers are biologically derived, while some have synthetic origin. The biodegradable fertilizers are eco-friendly and pose no threats to the environment. However, the synthetic fertilizers may remain in the soil or get discharged as effluents, causing environmental concerns.

Recently, there has been a trend to use semiconductor materials in agricultural applications. The optical property, among other properties of the semiconductors, has been identified as a suitable property to be utilized for agricultural applications. The optical properties of semiconductors are also influenced by certain factors. The sizes, shapes and surface area to volume ratio are some of the factors that control the optical properties of semiconductors [1]. The absorption of photon by a semiconductor leads to the excitation of an electron from the valence band to the conduction band (Figure 4.1). This leads to the generation of electron-hole pairs in the semiconductor system. The bandgap signifies the least energy required to form the charge carriers. When the absorbed energy is greater than or equal to the bandgap, the electron-hole pairs are created. This is called the optical transition.

The present trend is to explore the optical properties of the semiconductors in their nano-dimensions [2]. Reducing the particle size of the material to nano-range helps to manoeuvre the optical, physical and chemical properties at the molecular level. The use of nanomaterials helps to reduce the amount of application and also the nutrient losses in fertilizers. The semiconductor nanomaterials have been applied in agricultural applications in the form of fertilizers. Pesticide control and monitoring of pesticide residue have also been achieved with the help of sensors comprising semiconductors. Titanium dioxide is a semiconductor, which is widely known for its photocatalytic activity. This property has been utilized in various applications in the bulk as well as nano-form. It is sometimes used as a sensor, or as a part of fertilizer or sometimes modified to broaden the spectrum of its photocatalytic activity. There are similar semiconductors also

DOI: 10.1201/9781003188582-4

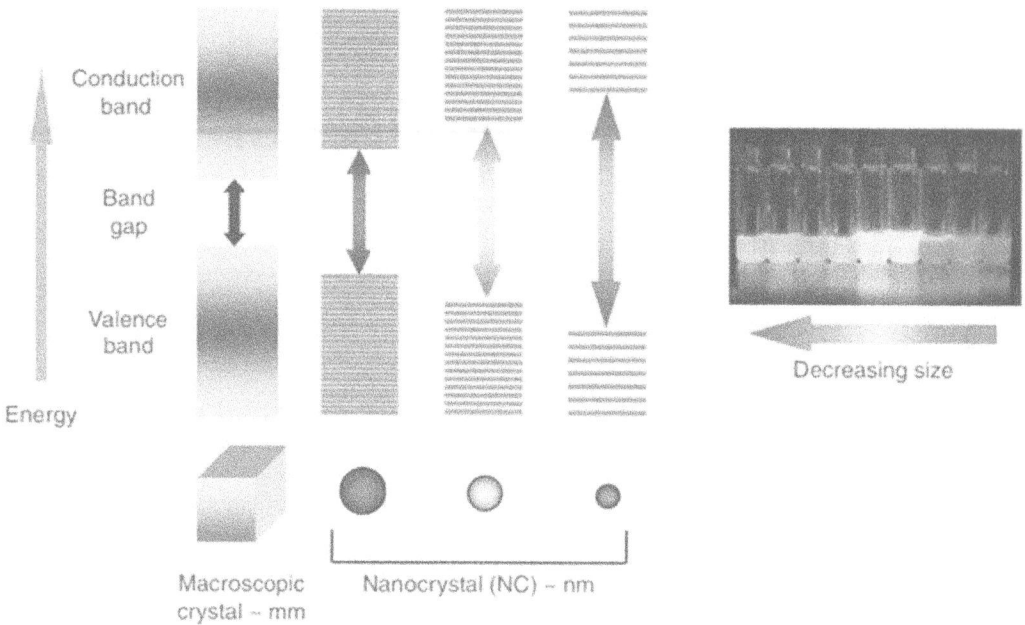

FIGURE 4.1 Bandgaps in semiconductors [1].

that have found application in agricultural uses. Silica nanomaterial is also being increasingly explored in agricultural fields.

This chapter discusses various semiconductors and their applications in the agricultural sectors. The past, present and future optical utilizations of the semiconductors have been discussed. The need for semiconductors and their advancements in the agricultural sector has also been highlighted.

4.2   SEMICONDUCTOR PHOTOCATALYST

The semiconductor photocatalysts are classified as oxide photocatalysts and non-oxide photocatalysts. The titanium-dioxide- and bismuth-oxide-based photocatalysts are some of the important oxide-based photocatalysts. TiO_2 exists in three crystal forms, rutile, anatase and brookite. Among these, only rutile and anatase forms show photocatalytic activity. TiO_2 has a wide bandgap energy of 3.2 eV and finds applications in ultraviolet (UV)-blocking purposes. There are also certain techniques followed to increase the photocatalytic efficiency of the photocatalysts (Figure 4.2). Bi_2O_3 photocatalyst exists in α, β, γ and δ crystal forms. Among these, the α form is the thermodynamically stable form of Bi_2O_3 and is most widely used as photocatalyst. It has a bandgap energy of 2.85 eV. This allows it to absorb light in the visible region of the spectrum [3, 4]. Transition metal oxides like ZnO, WO_3 and Fe_2O_3 are some of the other oxide photocatalysts that are in use. ZnO is known to have photocatalytic as well as piezoelectric properties. WO_3 with a bandgap energy of 208 eV can be used as a main and cocatalyst in many applications. It offers several advantages like large surface area and absorbing capabilities. It can also be used as an invisible material.

Some of the non-oxide photocatalysts include CdS, CuS, ZnS and C_3N_4. CdS is a semiconductor catalyst that has a bandgap energy of 2.42 eV and has been extensively studied as a photocatalyst. However, it gets corroded during application and causes issues relating to recovering of photocatalyst. This is one of the main limitations of CdS as a photocatalyst. It can be used as a photocatalyst in the range of visible light and up to a certain range in the UV spectrum. CuS photocatalyst has a bandgap energy of 2.2 eV and absorbs in the visible light region. It is one of the important metal sulphide semiconductors, which is widely used in optical, mechanical, electrical and sensor

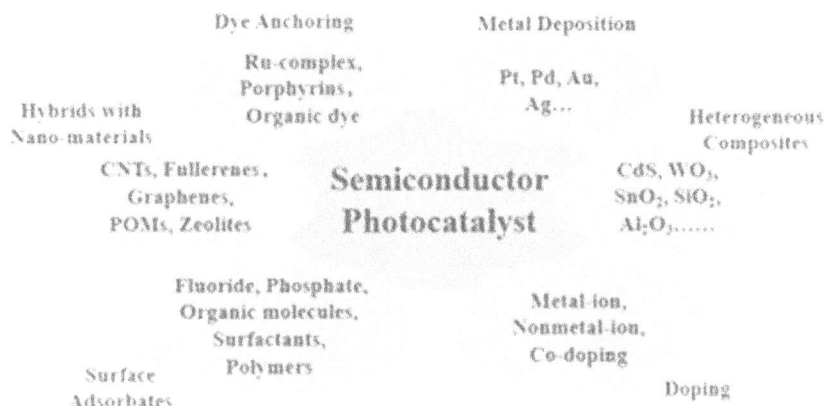

FIGURE 4.2 Techniques to improve photocatalytic efficiency [5].

utilizations. ZnS has bandgap energy between 3.6 eV and 3.8 eV. The nanomaterials of ZnS show good photocatalytic activity and are also non-toxic. They are available in the form of particles [6] and wires [7] of nano-dimensions that find a variety of applications.

4.3 SEMICONDUCTORS UTILIZED FOR AGRICULTURAL PURPOSE

4.3.1 TiO$_2$ BASED

Titanium dioxide (TiO$_2$) has distinguished applications in various fields because of its unique properties [8]. It has been explored as a photocatalyst, in dye-sensitized solar cells and even in biomedical devices. Agriculture is still a new field that requires further exploration of TiO$_2$. The suitable physical and chemical properties of the semiconductor TiO$_2$, its low cost and easy availability render it appropriate to be utilized in agricultural applications. The nanomaterials of TiO$_2$ have large bandgap energy of 3.2 eV, which permits them to function in the UV region of light. This limits their applications for agricultural purpose.

Pesticides are a necessary evil in agriculture. They are used widely in the agricultural fields; however, their excessive usage has an adverse effect both on the environment and the humans. It is therefore necessary to get rid of such toxicants. Semiconductor photocatalysis has been considered a suitable approach for pesticide degradation [9]. TiO$_2$ in the presence of light of energy, equal to or higher than its bandgap, leads to the generation of electron-hole pairs. In the presence of water, the semiconductor surface absorbs the hydroxyl ions. The holes react with these hydroxyl ions to generate hydroxyl radicals that are the strongest oxidants. These radicals oxidize the surface-adsorbed pesticides, converting them into biodegradable and non-toxic products like water and carbon dioxide. The detailed mechanism has been shown in Figure 4.3.

However, there are a few concerns in this process. The high rate of electron-hole pair recombinations and the high light energy required for the process affect the efficacy of the photocatalytic degradation of pesticides. As a solution to the problems, TiO$_2$ nanomaterials have been modified with other metal oxides [11]. On doping TiO$_2$ with WO$_3$, effective charge separation takes place due to the difference in their energy levels [12]. The doping carried out resulted in smaller clusters with increased surface area and hence better contact between the pesticides and the nanomaterials. Several doping and other techniques have been explored to increase the efficiency of the TiO$_2$ photocatalytic degradation of the pesticides [13].

The important ingredients for plant growth are oxygen and water intake. The intake of water and oxygen is facilitated by the superoxide and hydroxide ions produced during the photocatalytic process. Nano TiO$_2$ particles help in the absorption of inorganic nutrients that aid in seed germination

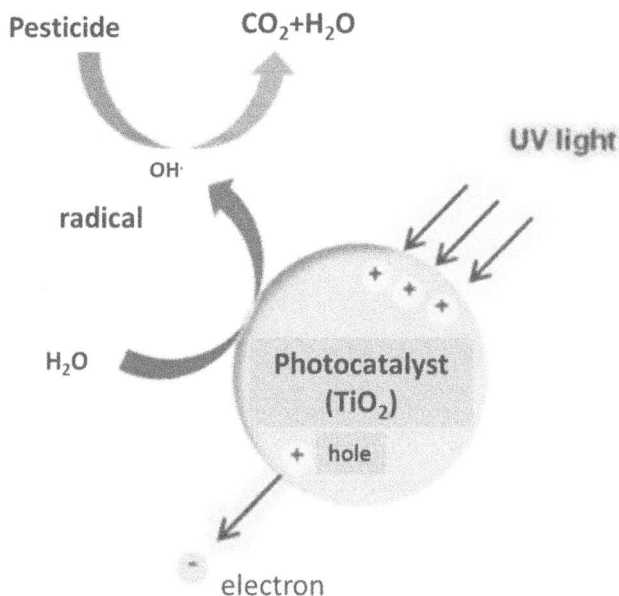

FIGURE 4.3 Photocatalytic degradation of pesticides in the presence of semiconductor TiO_2 [10].

and plant growth. In some cases, the greenery and freshness of leaves are retained due to better nitrogen fixation by the nano TiO_2 particles. Several other positive effects on plant growth have been observed in the presence of nanomaterials of TiO_2 [14, 15]. As stated earlier, during the photocatalytic process, by-products like active oxygen species are generated. These species have antimicrobial effect and hence offer an effective control of crop disease. This photocatalytic reaction utilizes the UV radiation. To improve the photocatalytic efficiency of TiO_2 in the visible region, it is doped with a photosensitive dye that allows the photocatalysis to take place in the visible region of spectrum. TiO_2 has also been used for water purification purpose. The reactive species may be generated at the surface of the nanomaterials, which on being UV radiated result in photocatalysis. The removal of organic pollutants and demineralization can be achieved through the photocatalytic process. The nanomaterials also are good detectors and are explored for sensor applications. These are used in pesticide residue detection because of their high sensitivity, high selectivity and quick response.

4.3.2 SILVER BASED

Silver nanoparticles have been known for their distinguished optical [16] and physical properties [17, 18]. Silver has therefore been explored for its application in various fields. It has been used as catalyst detector and sensor. It has extended applications in fields like optical biolabelling. Silver has antibacterial activity. This has been put to use in several forms like in pigments, in conductive/antistatic composites. It has also been used in biocide applications. Silver shows such properties both in nano-dimensions and in ionic bulk form. The use of silver nanomaterial is widely explored because of the high surface area offered by the nanomaterials. They have been explored in agriculture in the form of pesticide application. Silver in ionic form or of nano-dimensions has shown antimicrobial activity. Due to this property, they have been used to curb plant diseases. In one of the studies, silver ions were tested for their antifungal properties. They were tested against the phytopathogenic fungi, *Bipolaris sorokiniana* and *Magnaporthe grisea*. The silver ions were capable of reducing the colony formation of both fungal species; however, the extent of reduction was not the same. The antifungal activity of the silver ions is species specific, which is also evident from this study. *B. sorokiniana* had a higher tolerance to the silver ions compared to *M. grisea*, resulting in higher EC_{50} values of *B. sorokiniana* [19]. Silver nanomaterials have also been used as a resistive measure against the powdery mildew formation in roses. They caused a decline in the mildew formation at lower amount of application compared to the conventional fungicides [20]. The application of silver in agriculture is rather controversial due to the environmental issues associated with it. The silver ions or nanomaterials that are discharged in the sludge remain in the soil and affect the soil microbial biomass [21, 22].

4.3.3 SILICA/SILICON BASED

Silica is the second most abundant element after oxygen is available on the earth's surface [23]. It is a metalloid with appreciable physical and chemical properties. Silica has been used in biosensing and nano-bioimaging applications among other wide range of its utilization. Silica is an optically transparent material that allows the excited and emitted light to pass through when encapsulating an optically sensitive material. It also helps to improve the photostability of the compound [24]. Silicon is required to strengthen the cell walls of plants [25]. It has been used in bulk and as nanomaterials in agricultural applications. The nanomaterials show different properties than the bulk due to the difference in their size and the increased surface area to weight ratio. Silica nanoparticles (SiNPs) find application as nanoherbicides, nanofertilizers and also as nutrient delivery agents. They are also used to check the soil quality by monitoring the soil water retention capacity as nanosensors (Figure 4.4).

SiNPs have been used as pesticides. Their usage as pesticides is basically in two modes. They are sometimes directly applied to the fields for killing larvae of insects and others. In another mode, they are used as mesoporous SiNPs that release the effective pesticide gradually, acting as nanocarriers.

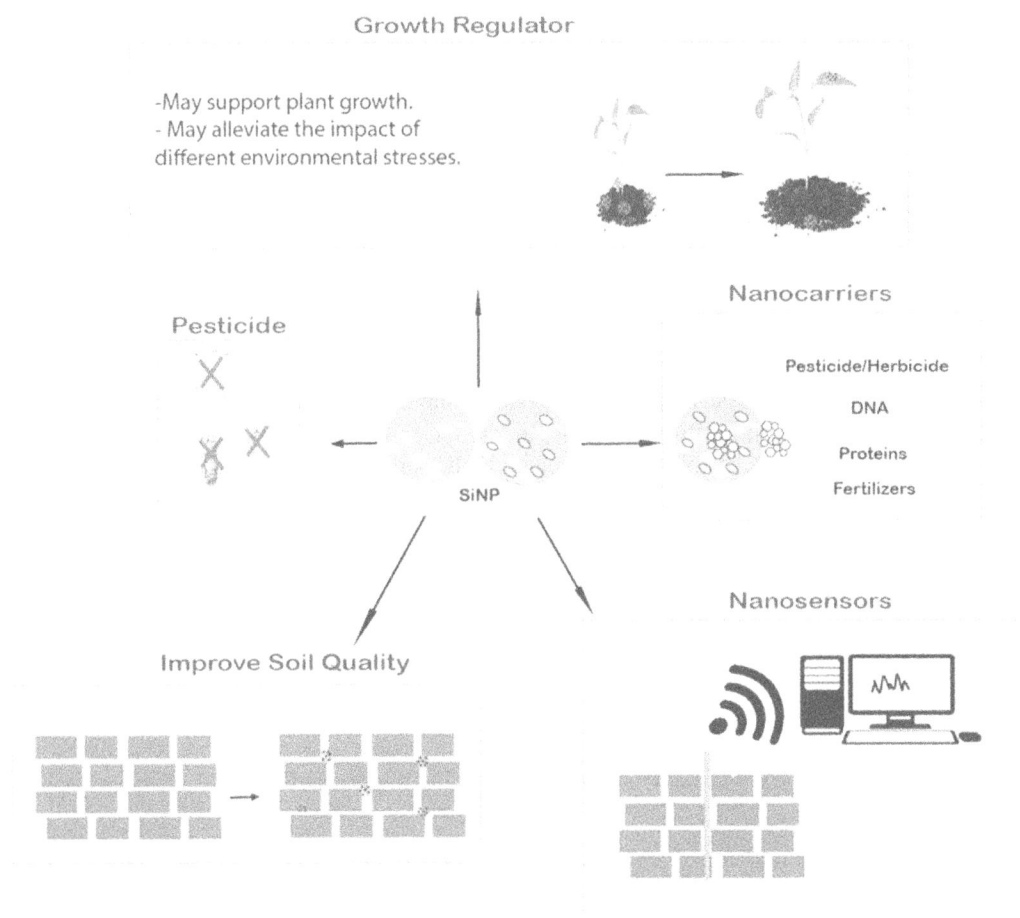

FIGURE 4.4 Use of silicon in agriculture [23].

These nanoparticles cause dehydration of the larvae leading to dysfunctional digestive tract or enlargement of the integument. In some cases, the nanomaterial leads to the blockage of the tracheas of the insects or destruction of the waxy coating due to abrasion. In order to achieve gradual release of the pesticides, mesoporous SiNPs are required [26]. The thickness of the shells also acts as a protective barrier against UV radiation. This also helps to increase the durability and efficiency of the pesticides used. They have also proved to be efficient target release agents. Such type of activity has been explored in fertilizers- and herbicides-containing SiNPs. They are also used as vectors for urea-based, boron-based and nitrogen-based fertilizers [27, 28]. In agriculture, soil also plays an important role in increasing the crop productivity. SiNPs used in nanozeolites help to increase the water-holding capacity of the soil. They act as slow-release agents of water to improve the water content in the soil. Some porous nanozeolites aid in water infiltration and retention owing to their capillary properties [29, 30]. SiNPs have also been incorporated in nanosensors for the detection of heavy metals in soils due to their superior optical sensitivity. The core-shell structure of the nanomaterials helps to improve the sensitivity and accuracy of the sensors. Even combinations of SiNPs with other nanomaterials have also been used for sensing purposes. One such instance is the combination of SiNPs and silver nanospheres for melamine detection. A ratiometric nanosensor was designed based on carbon dots. The surface of these dots was coated with rhodamine-B-doped SiNPs. Such types of sensors helped in the detection of copper (Cu^{2+}) ions. N-(β-aminoethyl)-γ-aminopropyl methyldimethoxysilane (AEAPMS) was used to synthesize the carbon dots. This resulted in residual ethylenediamine and methoxysilane groups on the carbon dots' surface that acted as recognition sites for the copper ions [31]. The mechanism has been shown in Figure 4.5.

4.3.4 ZINC OXIDE

Zinc oxide is a semiconductor that exhibits n-type conductivity. It also shows optical property that contributes to intrinsic as well as extrinsic effects. The intrinsic effect corresponds to the electron-hole pairs generated in the conduction and valence band. The extrinsic effects are the results of doping that generates defects. Zinc oxide has a large bandgap energy that allows it to be used for optical applications in the UV range [32]. This has been exploited in agricultural applications and used as UV-blocking agents. The photocatalytic property of zinc oxide is also well known. It absorbs a larger spectrum of the UV radiation compared to TiO_2. It may be represented in Figure 4.6.

Zinc-deficient soil is sometimes the cause of limited agricultural productivity. Therefore, zinc may be supplied in the form of ZnO nanoparticles or bulk form to restore the required zinc levels necessary for better agricultural yield. In cases where a small amount of zinc is required, it may be blended or coated onto some macro-fertilizers and supplied to the soil. However, the zinc penetration into the soil is dependent on the solubility of the source in the soil. To study this aspect, the dissolution

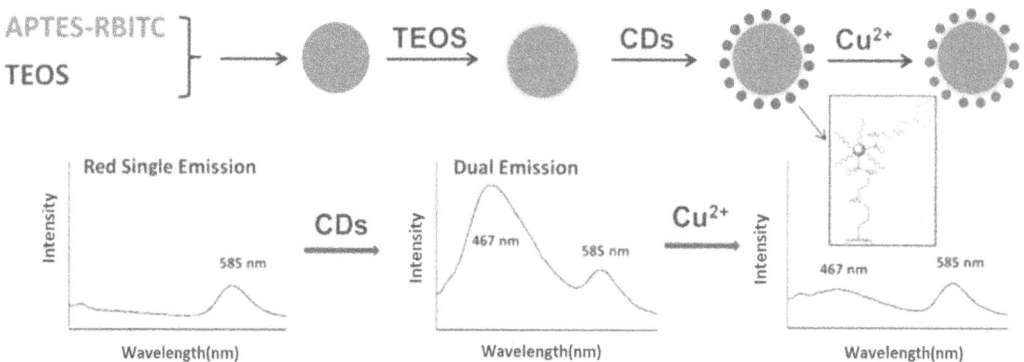

FIGURE 4.5   Mechanism of detection of Cu^{2+} ions [31].

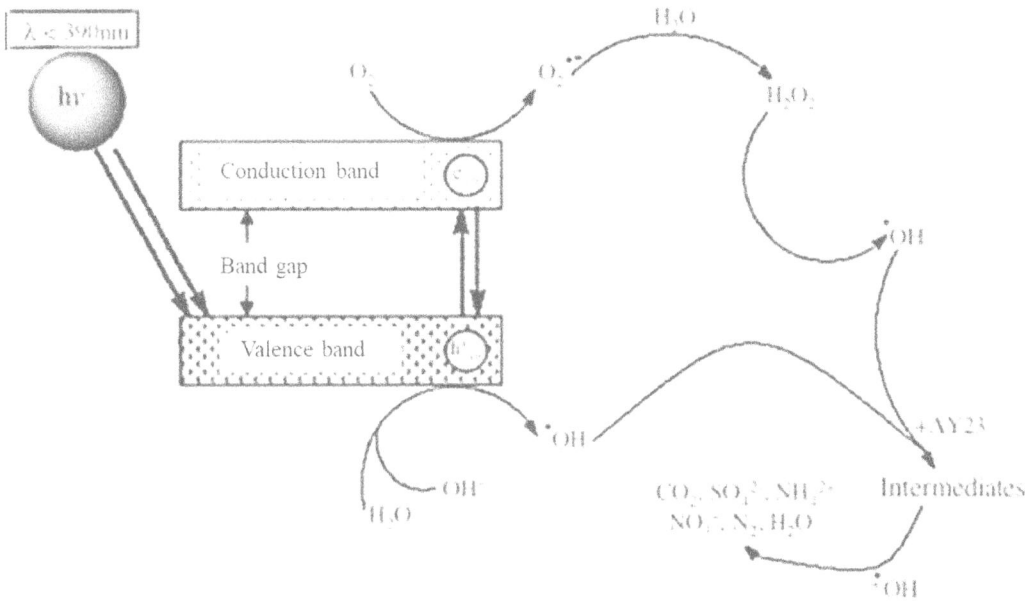

FIGURE 4.6 Mechanism of photocatalysis [33].

kinetics of fertilizer coated with zinc oxide nanoparticles was carried out by Milani *et al.* It was found that the ZnO nanoparticles had a tendency to agglomerate due to higher Brownian motion leading to more frequent collisions and hence the formation of aggregates. It was also observed that the ZnO-coated monoammonium phosphate (MAP) granules were better in releasing Zn compared to the ZnO-coated urea granules. This is due to their different level of pH in the aqueous solutions. The acidic pH of the ZnO-coated MAP granules in water caused higher dissolution of Zn. On the other hand, the alkalinity of the solution of ZnO-coated urea granules in water impaired the Zn dissolution. This also led to differences in the release rate of zinc [34]. The nitrogen cycle (Figure 4.7)

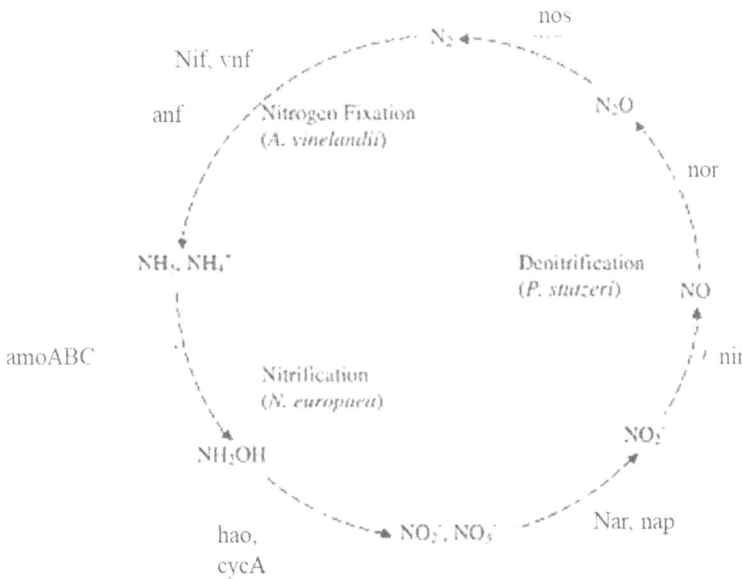

FIGURE 4.7 The nitrogen cycle and bacteria used by Yang *et al.* [35].

is also crucial to the soil fertility. Quantum dots beyond a definite concentration level are lethal to this cycle. Quantum dots comprising a metalloid crystallite core with an encapsulating ZnS or CdS shell material coated with organic molecules help to enhance the biocompatibility and stability of structure [35].

4.4 CONCLUSION

Semiconductors in their nano-dimensions or bulk form have found extensive applications in the field of agriculture. Significant areas of semiconductor utilization in agriculture include pesticide degradation, use in fertilizers, plant disease control, monitoring of residual pesticide content and enhancement of soil fertility. The photocatalytic property of the semiconductors is well utilized to their potential in such applications. The semiconductor catalysts can be categorized as oxide- and non-oxide-based. Both these categories of semiconductor photocatalysts have different bandgap energies that allow their implementation in different ranges of the light spectrum. Some are suitable for the visible range, while some are used in the UV region, acting as UV-blocking agents. Among the semiconductors, TiO_2 finds extensive applications in its nano-form and bulk form. The suitable physical and chemical properties of the semiconductor, along with its easy availability and cost-effectiveness, make TiO_2 the most preferred semiconductor photocatalyst. Other semiconductors that are used for agricultural purposes include silver, silica and zinc oxide.

There is a recent trend of research towards the application of semiconductor nanomaterials for agricultural purpose. It is due to the obvious advantages offered by the nanomaterials compared to the bulk form of semiconductor. However, there are some potential hazards that are offered by the nanomaterials towards the ecosystem, which should also be given sufficient consideration. This chapter gives good insight on the present status and the future prospects of semiconductor applications in agriculture. The optical properties of semiconductors that render them suitable for agricultural applications are also discussed with respect to different classes of semiconductor photocatalysts. The mechanism behind the use of these semiconductors in agricultural applications has also been explained. It has been observed that much remains to be done in this field and hence extensive research is being carried out to meet the demands of the agricultural productivity through semiconductor optical utilization.

REFERENCES

1. K.R. Choudhury, F. So, Z. Kafafi, Colloidal semiconductor nanocrystal-enabled organic/inorganic hybrid light emitting devices, Compr. Nanosci. Technol. 1–5 (2011) 183–214. https://doi.org/10.1016/B978-0-12-374396-1.00127-6
2. M.K. Nayak, J. Singh, B. Singh, S. Soni, V.S. Pandey, S. Tyagi, Introduction to semiconductor nanomaterial and its optical and electronics properties. In: Metal Semiconductor Core-Shell Nanostructures for Energy and Environmental Applications, Elsevier Inc., 2017. https://doi.org/10.1016/B978-0-323-44922-9.00001-6
3. X. Qian, D. Yue, Z. Tian, M. Reng, Y. Zhu, M. Kan, T. Zhang, Y. Zhao, Carbon quantum dots decorated Bi_2WO_6 nanocomposite with enhanced photocatalytic oxidation activity for VOCs, Appl. Catal. B Environ. 193 (2016) 16–21. https://doi.org/10.1016/j.apcatb.2016.04.009
4. K. Hu, L. E, Y. Li, X. Zhao, D. Zhao, W. Zhao, H. Rong, Photocatalytic degradation mechanism of the visible-light responsive $BiVO_4$/TiO_2 core–shell heterojunction photocatalyst, J. Inorg. Organomet. Polym. Mater. 30 (2020) 775–788. https://doi.org/10.1007/s10904-019-01217-w
5. F. Zhang, X. Wang, H. Liu, C. Liu, Y. Wan, Y. Long, Z. Cai, Recent advances and applications of semiconductor photocatalytic technology, Appl. Sci. 9 (2019). https://doi.org/10.3390/app9122489
6. R. Zhou, M.I. Guzman, CO_2 reduction under periodic illumination of ZnS, J. Phys. Chem. C. 118 (2014) 11649–11656. https://doi.org/10.1021/jp4126039
7. D.F. Moore, Y. Ding, Z.L. Wang, Crystal orientation-ordered ZnS nanowire bundles, J. Am. Chem. Soc. 126 (2004) 14372–14373. https://doi.org/10.1021/ja0451057

8. P. Roy, S. Berger, P. Schmuki, TiO_2 nanotubes: Synthesis and applications, Angew. Chemie – Int. Ed. 50 (2011) 2904–2939. https://doi.org/10.1002/anie.201001374

9. S. Ahmed, M.G. Rasul, R. Brown, M.A. Hashib, Influence of parameters on the heterogeneous photocatalytic degradation of pesticides and phenolic contaminants in wastewater: A short review, J. Environ. Manage. 92 (2011) 311–330. https://doi.org/10.1016/j.jenvman.2010.08.028

10. Y. Wang, C. Sun, X. Zhao, B. Cui, Z. Zeng, A. Wang, G. Liu, H. Cui, The application of nano-TiO_2 photo semiconductors in agriculture, Nanoscale Res. Lett. 11 (2016) 1–7. https://doi.org/10.1186/s11671-016-1721-1

11. X. Zhang, F. Wu, Z. Wang, Y. Guo, N. Deng, Photocatalytic degradation of 4,4′-biphenol in TiO_2 suspension in the presence of cyclodextrins: A trinity integrated mechanism, J. Mol. Catal. A Chem. 301 (2009) 134–139. https://doi.org/10.1016/j.molcata.2008.11.022

12. N.A. Ramos-Delgado, M.A. Gracia-Pinilla, L. Maya-Treviño, L. Hinojosa-Reyes, J.L. Guzman-Mar, A. Hernández-Ramírez, Solar photocatalytic activity of TiO_2 modified with WO_3 on the degradation of an organophosphorus pesticide, J. Hazard. Mater. 263 (2013) 36–44. https://doi.org/10.1016/j.jhazmat.2013.07.058

13. H. Guan, D. Chi, J. Yu, X. Li, A novel photodegradable insecticide: Preparation, characterization and properties evaluation of nano-Imidacloprid, Pestic. Biochem. Physiol. 92 (2008) 83–91. https://doi.org/10.1016/j.pestbp.2008.06.008

14. L. Zheng, F. Hong, S. Lu, C. Liu, Effect of nano-TiO_2 on strength of naturally aged seeds and growth of spinach, Biol. Trace Elem. Res. 104 (2005) 83–91. https://doi.org/10.1385/bter:104:1:083

15. G. Song, Y. Gao, H. Wu, H. Hou, C. Zhang, H. Ma, Physiological effect of anatase TiO_2 nanoparticles on Lemna minor, Environ. Toxicol. Chem. 31 (2012) 2147–2152. https://doi.org/10.1002/etc.1933

16. D.D. Jr Evanoff, G. Chumanov, Synthesis and optical properties of silver nanoparticles and arrays, 29634 (2005) 1221–1231. https://doi.org/10.1002/cphc.200500113

17. I.A. Wani, S. Khatoon, A. Ganguly, J. Ahmed, A.K. Ganguli, T. Ahmad, Silver nanoparticles: Large scale solvothermal synthesis and optical properties, Mater. Res. Bull. 45 (2010) 1033–1038. https://doi.org/10.1016/j.materresbull.2010.03.028

18. C. Baker, A. Pradhan, L. Pakstis, D.J. Pochan, S.I. Shah, Synthesis and antibacterial properties of silver nanoparticles, J. Nanosci. Nanotechnol. 5 (2005) 244–249. https://doi.org/10.1166/jnn.2005.034

19. Y.K. Jo, B.H. Kim, G. Jung, Antifungal activity of silver ions and nanoparticles on phytopathogenic fungi, Plant Dis. 93 (2009) 1037–1043. https://doi.org/10.1094/PDIS-93-10-1037

20. A. Gogos, K. Knauer, T.D. Bucheli, Nanomaterials in plant protection and fertilization: Current state, foreseen applications, and research priorities, J. Agric. Food. Chem. 60 (2012) 9781–9792.

21. J. Liu, H.H. Robert, Ion release kinetics and particle persistence in aqueous nano-silver colloids, Environ. Sci. Technol. 44 (2010) 2169–2175. https://doi.org/10.1021/es9035557

22. A.M.E.L. Badawy, R.G. Silva, B. Morris, K.G. Scheckel, M.T. Suidan, Surface charge-dependent toxicity of silver nanoparticles, Environ. Sci. Technol. 45 (2011) 283–287.

23. A. Rastogi, D.K. Tripathi, S. Yadav, D.K. Chauhan, M. Živčák, M. Ghorbanpour, N.I. El-Sheery, M. Brestic, Application of silicon nanoparticles in agriculture, 3 Biotech. 9 (2019). https://doi.org/10.1007/s13205-019-1626-7

24. P. Tallury, K. Payton, S. Santra, Silica-based multimodal/multifunctional nanoparticles for bioimaging and biosensing applications, Nanomedicine. 3 (2008) 579–592. https://doi.org/10.2217/17435889.3.4.579

25. E. Epstein, The anomaly of silicon in plant biology, Proc. Natl. Acad. Sci. U.S.A. 91 (1994) 11–17. https://doi.org/10.1073/pnas.91.1.11

26. I.I. Slowing, J.L. Vivero-Escoto, C.W. Wu, V.S.Y. Lin, Mesoporous silica nanoparticles as controlled release drug delivery and gene transfection carriers, Adv. Drug Deliv. Rev. 60 (2008) 1278–1288. https://doi.org/10.1016/j.addr.2008.03.012

27. F. Torney, B.G. Trewyn, V.S.Y. Lin, K. Wang, Mesoporous silica nanoparticles deliver DNA and chemicals into plants, Nat. Nanotechnol. 2 (2007) 295–300. https://doi.org/10.1038/nnano.2007.108

28. H. Wanyika, E. Gatebe, P. Kioni, Z. Tang, Y. Gao, Mesoporous silica nanoparticles carrier for urea: Potential applications in agrochemical delivery systems, J. Nanosci. Nanotechnol. 12 (2012) 2221–2228. https://doi.org/10.1166/jnn.2012.5801

29. J. Szerement, A. Ambrożewicz-Nita, K. Kędziora, J. Piasek, Use of zeolite in agriculture and environmental protection. A short review, Dep. Phys. Chem. Porous Mater. 781 (2014) 172–177.

30. S.N. Abdel-Hassan, A.M. Abdullah Radi, Effect of zeolite on some physical properties of wheat plant growth (*Triticum aestivum* L.), Plant Arch. 18 (2018) 2641–2648.

31. X. Liu, N. Zhang, T. Bing, D. Shangguan, Carbon dots based dual-emission silica nanoparticles as a ratiometric nanosensor for Cu2+, Anal. Chem. 86 (2014) 2289–2296.
32. J. Wu, D. Xue, Progress of science and technology of ZnO as advanced material, Sci. Adv. Mater. 3 (2011) 127–149. https://doi.org/10.1166/sam.2011.1144
33. M.A. Behnajady, N. Modirshahla, R. Hamzavi, Kinetic study on photocatalytic degradation of C.I. Acid Yellow 23 by ZnO photocatalyst, J. Hazard. Mater. 133 (2006) 226–232. https://doi.org/10.1016/j.jhazmat.2005.10.022
34. N. Milani, M.J. McLaughlin, S.P. Stacey, J.K. Kirby, G.M. Hettiarachchi, D.G. Beak, G. Cornelis, Dissolution kinetics of macronutrient fertilizers coated with manufactured zinc oxide nanoparticles, J. Agric. Food Chem. 60 (2012) 3991–3998. https://doi.org/10.1021/jf205191y
35. Y. Yang, J. Wang, H. Zhu, V.L. Colvin, P.J. Alvarez, Relative susceptibility and transcriptional response of nitrogen cycling bacteria to quantum dots, Environ. Sci. Technol. 46 (2012) 3433–3441. https://doi.org/10.1021/es203485f

5 Nonlinear Optical Properties of Semiconductors, Principles, and Applications

Muhammad Rizwan, Aleena Shoukat, Asma Ayub,
Iqra Ilyas, Ambreen Usman, and Seerat Fatima

CONTENTS

5.1 INTRODUCTION

Semiconductors (SCs) have extensive applications in the field of electronics, in which their electronic properties have been manipulated via doping for technological applications. SCs have been utilized widely in the field of optoelectronics, so their optical properties are also of interest. Linear optical properties of SCs have been extensively reviewed. In this chapter, focus will be over the nonlinear optical properties of SCs. Inquisition of nonlinear optical properties through theoretical and experimental techniques has been at large in research field. Nonlinear optical properties of SCs have been correlated with lasers due to high excitation phenomena.

Nonlinear optical properties are the response of SC materials to high-intensity field just as observed in lasers. Nonlinear optical properties are of prodigious interest from the point of view of basic physics and potential application of nonlinear effects such as phase conjugation, bistability in electronics or optoelectronics (Haug and Schmitt-Rink 1984). The evolution in nonlinear optical

DOI: 10.1201/9781003188582-5

49

properties is directly correlated with the progress of lasers and their properties. Investigation and exploration of resonant excitation were made possible due to tunable lasers; similarly, picosecond and femtosecond (fs) lasers enabled us to study nonlinear optical effects such as interband relaxation, interband combination, and phase relaxation. Nonlinear optical properties of SCs have garnered massive attention due to their large magnitude near band energies (Hönerlage et al. 1985).

It is hard to contain all information about nonlinear optical properties in one chapter. So, we will be focusing on the main aspects of nonlinear optical properties and their applications (Klingshirn and Haug 1981). For optical properties, optical parameters such as reflection via reflectivity coefficient $R(\omega)$, transmittance via transmission coefficient $T(\omega)$, and luminescence $L(\omega)$, by complex dielectric function and complex refractive index (RI), have been considered. Absorption coefficient $\alpha(\omega)$ was correlated with $R(\omega)$, $T(\omega)$, $n(\omega)$, and $\varepsilon(\omega)$ (Klingshirn 1990).

In linear optics, we can easily exploit the dependence of $\varepsilon(\omega)$ on the frequency ω. Frequency dependence is studied through investigation of resonances with which electromagnetic field can successfully be coupled. In SCs these resonances are optical phonon modes or plasmon (Schmitt-Rink, Chemla, and Miller 1989). In the region of absorption edge, the common resonance observed is band-to-band transitions. There is an assumption made in linear optics that optical properties are independent of intensity of light and polarization $P(\omega)$ of the medium oscillates with equal frequency as the incident light (Ironside 1993). In linear optics, it is assumed that two light beams that interconnect in the medium don't interact with each other.

When nonlinear optics (NLO) is in sight, the point of view changes and all properties now depend on the intensity of light. The intensity-dependent optical parameters are due to third-order optical nonlinearities. Third-order nonlinearities are further divided into resonant and nonresonant optical nonlinearities, which will be discussed in detail in the next sections. Resonant nonlinearities are due to photo-generated carriers and energies that are very close to the absorption edge. Nonresonant nonlinearities occur at energies lower than absorption edge. When discussing resonant nonlinearities, it is more convenient to refer them as intensity-dependent optical properties. The most important consequence of resonant effects is that it gives insight into the dynamics of free carriers in SCs. Nonlinear theory explains third-order nonlinearity.

5.2   NONLINEAR OPTICS

When light interacts with matter, multiple processes such as absorption, reflection, and luminescence can occur. All these phenomena are rendered as optical properties of a material but are dependent on wavelength and not on the intensity of light. But at a higher intensity of light, these properties can be altered by electromagnetic field. These intensity-dependent optical properties are of great significance and have been studied via NLO (Drake 2006).

Nonlinear optics, which is inevitable in order to understand nonlinear optical properties and effects, is a complete material phenomenon and is often referred to as NLO. Maxwell equations based on interaction of light with matter provide all the information about laws of optics. These equations are applicable to all materials but depend on the characteristics of materials. Maxwell equations relate electric field (EF) **E** and current density, **j** via conductivity and the electric displacement **D** with **E** via permittivity ε. These relations are given in the following equation:

$$J = \sigma E \tag{5.1a}$$

where j is current density, E is electric field, and σ is conductivity.

$$\mathbf{D} = \varepsilon\mathbf{E} \tag{5.1b}$$

where ε is permittivity tensor, which can be written as

$$\varepsilon = \varepsilon_0\varepsilon_r \tag{5.2}$$

ε_r is relative permittivity and ε_o is the permittivity of free space.

Electric displacement can also be written as given in the following equation:

$$\mathbf{D} = \varepsilon_o \mathbf{E} + \mathbf{P} \tag{5.3}$$

where $\mathbf{P}$ is polarization that is equal to the number of dipole moments per unit volume plus gradients of multipole moments, but higher terms are neglected here. From Equation (5.3), we can write

$$\mathbf{P} = \varepsilon_o(\varepsilon_r - 1)\mathbf{P} = \varepsilon_o \chi \cdot \mathbf{E} \tag{5.4}$$

where χ is susceptibility tensor.

In the case of linear materials, the response in the material corresponds directly to its respective stimulus, for example, polarization is proportional to EF as in Equation (5.4), but susceptibility is independent of the field. This direct dependence between response and stimulus holds only at low fields, but at higher value of field polarization dependence on field is no longer applicable and now susceptibility becomes dependent on field, then this is called a nonlinear response. Nonlinear response is usually manifested as a deviation from the linear response, polarization then can be represented as a power series such as given in the following equation:

$$\frac{\mathbf{P}}{\varepsilon_o} = \chi^1 \cdot \mathbf{E} + \chi^2 : \mathbf{EE} + \chi^3 \cdot \mathbf{EEE} + \dots \tag{5.5}$$

where χ^1 represents the linear term called linear susceptibility tensor, χ^2 is the quadratic susceptibility and represents second harmonic generation (SHG) and frequency generation, and similarly, the third term represents cubic susceptibility tensor and represents wave mixing. Usually, third term is not required.

So, polarization $\mathbf{P}$ has nonlinear dependence on field for static EFs and can be written as

$$\mathbf{P}_{ind} = \mathbf{P}^1_{(\omega,k)} + \mathbf{P}^2_{(\omega,k)} + \mathbf{P}^3_{(\omega,k)} + \dots \tag{5.6}$$

$$\mathbf{P}^n = \chi^n_{\left(\omega = \omega^1 + \dots + \omega_n, k = k_1 + \dots + k_n\right)} \mathbf{E}_{\left(\omega^1, k_1\right)} \cdots \mathbf{E}_{\left(+\omega_n, k_n\right)} \tag{5.7}$$

where the nth term of polarization represents the interaction of n+1 EF and conservation of momentum and energy is valid. The linear term in polarization expression can have the same frequency as the incident EF, but nonlinear term includes combinations of frequencies, since two or more oscillating EFs give rise to nonlinear term as shown by higher order terms in Equation (5.5) (Munn and Ironside 1993a).

Nonlinear response includes the superposition of fields having different frequencies with an additional term of simultaneous response, so general linear superposition does not hold. Expanding equation, we can write

$$\mathbf{P}\left(\frac{\omega_o}{\varepsilon_o}\right) = \chi^1\left(-\omega_o; \omega_o\right) \cdot \mathbf{E}\left(\omega_o\right) + \chi^2\left(-\omega_o; \omega_1, \omega_2\right) : \mathbf{E}\left(\omega_1\right)\mathbf{E}\left(\omega_2\right)$$

$$+ \chi^3\left(-\omega_o; \omega_1, \omega_2, \omega_3\right) \cdot \mathbf{E}\left(\omega_1\right)\mathbf{E}\left(\omega_2\right)\mathbf{E}\left(\omega_3\right) + \dots \tag{5.8}$$

where ω_o is the output frequency and is given a negative sign; however, input frequency can be positive or negative. The sum of frequencies should be zero such that

$$\omega_o = \omega_1 + \omega_2 \tag{5.9}$$

We can write Equation (5.8) as for Cartesian coordinates.

$$P_\alpha \left(\frac{\omega_o}{\varepsilon_o} \right) = \chi \beta_\alpha \left(-\omega_o; \omega_o \right) \cdot \mathbf{E}_\beta \left(\omega_o \right) + \chi_{\alpha\beta\gamma} \left(-\omega_o; \omega_1, \omega_2 \right) : \mathbf{E}_\beta \left(\omega_1 \right) \mathbf{E}_\gamma \left(\omega_2 \right)$$

$$+ \chi_{\alpha\beta\gamma} \left(-\omega_o; \omega_1, \omega_2, \omega_3 \right) \cdot \mathbf{E}_\beta \left(\omega_1 \right) \mathbf{E}_\gamma \left(\omega_2 \right) \mathbf{E}_\delta \left(\omega_3 \right) + \dots \qquad (5.10)$$

where the subscripts α, β, γ, δ are Cartesian components.

If there is a single frequency, the response will be at fundamental frequency $\left(\omega_o \right)$ or higher frequencies that give rise to harmonic generation. For more than one frequency, there can be either sum frequency generation (SFG) or difference frequency generation (DFG) often referred to as parametric mixing. Positive frequencies of the EF refer to absorption, whereas negative ones refer to emission.

If some frequencies are the same, then not all susceptibility components are unique, since frequencies can be interchanged. The first term in Equation (5.10) is always symmetric under an interchange. In another case if all the frequencies are zero, then materials will have the Kleinman symmetry given that its frequencies are different from absorption frequency (Kleinman 1962a).

In this case, all susceptibility components are symmetric under any interchange. Symmetry of materials decides independent nonzero susceptibility components. All SCs possess nonlinear effects that become evident only at high values of intensities. The value of nonlinear response is a characteristic of a material. Nonlinear susceptibility is a tensor and its symmetry will reduce the number of tensor elements. For example, in centrosymmetric materials the second term in Equation (5.10) is always zero but third term is always nonzero. In materials, nonlinear response will manifest itself as intensity-dependent refractive index (IDRI) or absorption. In bulk SCs, nonlinear response arises due to thermal, electronic, or excitonic effects. Particular contribution of a different phenomenon to nonlinear effects is the function of materials, the intensity, and energy of light (Rohleder and Munn 1992).

5.3 GENERAL FORMULATION

Due to extensive work in the field of NLO, it is possible to generate a general formulation to explain what happens in an SC at high intensities due to real or virtual production of excitons or other nonlinear effects such as photothermal optical nonlinearities. Intensity-dependent nonlinear effects and creation of real or virtual excitons that give rise to optical nonlinearities are dependent on the nature of material, but for SCs we can characterize them into three regimes as discussed in the following.

In the first regime, that is, the linear or low excitation regime, the optical properties are decided by excitons or individual electron-hole pairs. These excitons can move freely in the material as a Bloch wave or can be neutral donors by being bound to defects present in the material at low temperatures. Excitons in linear regime don't interact with each other due to their low density. The only interaction that can ensue either between excitons or between excitons and free carrier is elastic or inelastic scattering. Scattering between excitons and free carrier can happen only at high temperature, where some excitons are thermally dissociated. There is an extra density-dependent increment in resonances of excitons due to these scattering that reduces the phase relaxation time and results in extra bands in absorption and luminescence spectra (Klingshirn and Haug 1981).

Second mechanism that leads to optical nonlinearities is due to biexciton. Biexcitons are two bound electron-hole pairs, whose formation and decay give rise to many nonlinear optical phenomena (Henneberger and Haug 1988; Klingshirn and Haug 1981).

Induced absorption resonance appears at $\hbar\omega_{ind}$, given that excitons are present at density N_{exc}:

$$\hbar\omega_{ind} = E_{biexc} - E_{exc} \qquad (5.11)$$

Oscillator strength of this resonance depends on N_{exc}.

Two-polariton absorption resonance $\hbar\omega_l$ occurs at energy E given as

$$\hbar\omega_{TPA} = E_{biexc} - \hbar\omega_l \tag{5.12}$$

The oscillator strength of this resonance is proportional to $\hbar\omega_l$. The decay of biexcitons gives new bands in absorption and luminescence spectra. Excitons can be created through real or virtual excitons, if biexcitons are created virtually by two photons $\hbar\omega_{exc}$, they can decay by a two-photon process called hyper-Raman process. In this process, the biexcitons simultaneously decay into an exciton and a quanta $\hbar\omega_K$:

$$\hbar\omega_R = 2\hbar\omega_{exc} - E_{exc} \tag{5.13}$$

Biexcitons can also decay via a stimulated process called stimulated hyper-Raman process or degenerate four waves mixing as explained by third term in Equation (5.5) (Hönerlage et al. 1985).

The optical stark effect is another process that occurs in this regime. In this case, an incident radiation field having frequency ω_p interacts with an exciton having frequency $\hbar\omega_{exc}$ (Koch et al. 1989; Schmitt-Rink, Chemla, and Miller 1989).

This interaction leads to a level of repulsion, thus a blue shift in excitons resonance. This phenomenon is referred to as optical stark effect produced due to incident AC EF beam. AC stark effect is preferably studied with short pulses in picosecond (ps) regime, because of high intensities required for it and blue shift of excitons.

In a high-density regime, electron-hole pairs are formed in high concentration that led to a mutual distance that is comparable to excitonic Bohr radius. In this regime, excitons are treated as electron-hole plasma (EHP) since the concept of excitons as individual quasiparticle becomes unacceptable. Formation of EHP greatly affects optical properties of SCs. With cumulating electron-hole pair density, the bandgap of SC decreases. This relationship between bandgap and carrier density holds in some covalent SCs, but there are some discrepancies in others. As the electron-hole density increases, Coulomb screening increases and this compensates the bandgap reduction for excitons resonance energies near K = 0 (Zimmermann 1988). This effect is valid to a few meVs in bulk and two-dimensional (2D) SCs. But there is a small shift in the case of 2D SCs due to less screening. At a certain value of electron hole density, called Mott density, the exciton-binding energy becomes zero (Hulin et al. 1986).

This density separates excitons in gaseous state from EHP state. Excitons resonances disappear for a higher value of electron-hole pair density due to effects such as Columbic screening and phase space filling, that is, the occupation of states near the absorption edge. This occupation of states is necessary for excitonic wave function and exchange interaction (Schmitt-Rink, Chemla, and Miller 1985). If the photon interacting with SC is from near fundamental absorption edge, it will create real or virtual electron-hole pairs (excitons or polarizations). In most SCs, recombination happens through non-radiative transition; thus, most incident light that is absorbed is transmuted into heat. If the quantity of heat produced is adequate, it will alter optical properties of SCs. More complicated nonlinear effect such as increment or decrement of absorption coefficient with respect to photon energy with increasing intensity can happen in SCs, if bandgap upsurges with temperature or the bound excitation absorption line shifts over the photon energy (Hellwege 1982; Klingshirn 1990).

5.4 HARMONIC GENERATION

Considering special cases of nonlinear effects in SCs, we take a monochromatic incident light that interacts with the SC. From Equation (5.6), we know that resonance occurs at fundamental frequency and there is a higher order of harmonics for frequencies 2ω, 3ω, etc. In Equation (5.6), the second term, called quadratic susceptibility $\chi^2(-2\omega; \omega, \omega)$ is responsible for the harmonic generation of second order in non-centrosymmetric materials. Similarly, the cubic susceptibility generates the third-order harmonic generation (Munn and Ironside 1993b).

Harmonic generation with respect to susceptibility order can also be considered frequency doubling up and tripling up, and so on. Frequencies of all inputs are the same, so susceptibility components of all order satisfy the symmetry property under any interchange except the first term in Equation (5.6), as given in the following two equations:

$$\chi_{\alpha\beta\gamma}\left(-2\omega; \omega, \omega\right) = \chi_{\alpha\gamma\beta\beta}\left(-2\omega; \omega, \omega\right) \tag{5.14}$$

$$\chi_{\alpha\beta\gamma\delta}\left(-3\omega; \omega, \omega, \omega\right) = \chi_{\alpha\beta\delta\gamma}\left(-3\omega; \omega, \omega, \omega\right) = \chi_{\alpha\gamma\beta\delta}\left(-3\omega; \omega, \omega, \omega\right) \tag{5.15}$$

Symmetry properties given in Equations (5.14) and (5.15) indicate that only six total combinations are possible for the Cartesian components $\beta\gamma$ instead of nine. To correlate polarization amplitude and optical EF, the subscripts $\beta\gamma$ are replaced with a single coordinate λ in SHG coefficient $d_{\alpha\lambda}$. For a particular experiment, we can write SHG coefficient as

$$d_{\text{eff}} = \frac{1}{2} \mathbf{e_p} \cdot \chi : \mathbf{e_1} \mathbf{e_2} \tag{5.16}$$

where $\mathbf{e_p}$ is the polarization vector and $\mathbf{e_1}$ and $\mathbf{e_2}$ are vectors of incident EF and are two vectors even though they come from the same incident field.

Second-order harmonic generation happens inside the medium, so corresponding polarization resides inside the material. In the case of isotropic and low symmetrical materials, the two vectors, $\mathbf{e_1}$ and $\mathbf{e_2}$, may be the same because the incident plane polarized light travels as two orthogonal waves or may be different and each vector corresponds to individual wave.

In anisotropic materials, polarizations are determined by electric displacement $\mathbf{D}$ plane instead of EF $\mathbf{E}$. Linear optics plays a huge role in determining the efficiency of second-order harmonic generation (Landau et al. 2013).

Second-order harmonic wave is generated by a fundamental wave, and its phase at any point in the medium is determined by the fundamental wave present at that particular point; there may be other second harmonic waves whose phases are determined by other fundamental waves, then entire second harmonic wave will be the result of the superposition of all the waves that have traveled along the path. Because of the phase difference between the second harmonic wave and fundamental wave, the intensity of the final net second harmonic wave obtained after superposition will vary periodically with the propagation direction. The reason for this relation is that both fundamental and second harmonic waves have different speeds that depend on the RI ($v = c/n$). When all the waves have traveled an integral number of their speed, they will interfere constructively and the intensity of the net second harmonic wave will be maximum. The distance traveled by waves is given as

$$K = 2\pi/\lambda = \omega n/c \tag{5.17}$$

where n and λ are RI and wavelength at a frequency ω in the material. The extreme intensity occurs when waves are coherent in length as

$$L_{\text{coh}} = \pi/\Delta k \tag{5.18}$$

For effectual second-order harmonic generation, phase matching and wave vectors should be matched for the beam. Phase matching refers to the same phase of harmonic wave and fundamental wave (Hobden 1967).

Phase matching is dependent on the frequencies of the fundamental and harmonic waves and the direction of incident light. There are two types of phase matching, one is type I phase matching that is possible for fundamental and harmonic waves that are collinear. When an ordinary or extraordinary ray is produced by two ordinary or extraordinary waves at the harmonic frequency, this satisfies the condition for index matching or type I phase matching and is given as

$$n_{2\omega} = n_{\omega} \qquad (5.19)$$

Second kind of phase matching is type II, in which the combination of one ordinary and one extraordinary wave at harmonic frequency produces an ordinary or extraordinary wave at harmonic frequency, this condition is given as

$$2n_{2\omega} = n_{\omega}^o + n_{\omega}^e \qquad (5.20)$$

where the superscripts o and e denote the ordinary and extraordinary wave refractive indices, respectively.

If these phase matching are not attained, intensity can be maximized by quasi-phase matching. In quasi-phase matching materials with thickness equal to coherence length are apart from each other by inactive regions that have the same thickness. In quasi-phase matching destructive interference is achieved in these inactive regions. Thus, only the active region adds up and gives rise to a stepwise increase in the net intensity. Optical rectification (OR) is due to zeroth harmonic or a constant field. OR is not prominent in nonlinear effects, but it is important for comprehension of high-order effects (Munn and Ironside 1993a).

Effects due to second- and third-order harmonic generations will be discussed in the following sections.

5.4.1 Second Harmonic Generation

SHG is an arising distinctive appliance for biotic visualization. In 1986, SHG was for the first time applied to the biotic visualization by Freund, Deutsch, and Sprecher (1986). To consider the polarity of the collagen filaments at low spatial resolution, in rodent tail ligament, the SHG imaging technique was applied. From that point onward, a few mechanical improvements were introduced that have changed SHG system being appropriate for collagen filaments imagery in organs like kidney, liver, and lungs just as in associating tissues like ligament, bones, skin, and veins (Kumar et al. 2015).

From the consequences of the addition of light fields and from phase matching, which are brought through the molecular adjustment having systematic non-centrosymmetric composition, the SHG is considered the second-order nonlinear coherent scattering technique. The two photons of an input laser that are at optical frequency ω are interceded by two fundamental conditions of the framework and these are also up changed above to a single photon on optical frequency 2ω; it has occurred in SHG cycle (Figure 5.1(a)). The SHG interaction begins from stimulated polarization instead of actual absorption (Kumar et al. 2012):

$$I_{BD} = 2\alpha_S \left[\chi_{1122}^{(3)\ R,re} - \chi_{1212}^{(3)\ R,re} \right] I_p I_s L \qquad (5.21)$$

$$I_{BD} = 2\alpha_S \left[\chi_{1122}^{(3)\ R,im} - \chi_{1212}^{(3)\ R,im} \right] I_p I_s L \qquad (5.22)$$

For SHG procedure, the second-order term of induced polarization is accountable and it may be resulting from the generic appearance of P^2 that is written as

$$P_{SHG}^{(2)} = 2\varepsilon_o \chi^{(2)} \left(-2\omega; \omega, \omega \right) \cdot A_\omega A_\omega \exp\left(j2\omega t \right) \qquad (5.23)$$

In the previous equation, $\chi^{(2)}$ is responsible for SHG and is known as an element of second-order susceptibility tensor. The $\chi^{(3)}$ retains some birefringent characteristics, the full $\chi^{(2)}$ tensor shares the

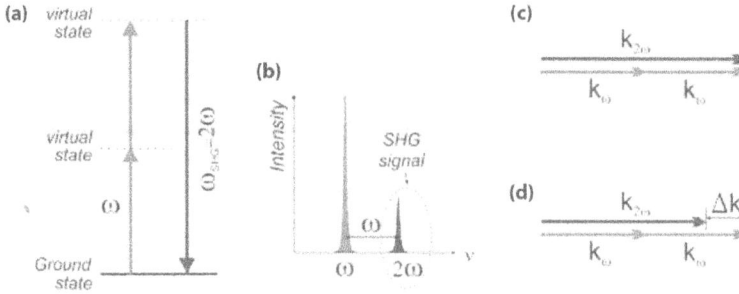

FIGURE 5.1   SHG procedure: (a) energy level diagram, (b) shows input and output field's spectral locations, (c) perfect synchronization of phase, and (d) normal wave-vector incompatibility in tissue surroundings that is remunerated by scattering and randomness (Kumar, Coluccelli, and Polli 2018).

similar type of these properties, and it has $3^3 = 27$ matrix elements. The equation for field propagation for $P^2 = p^2 \exp[j(2\omega)t]$ can be attained by way of the following:

$$\frac{\partial A_{2\omega}}{\partial z} = -j\frac{\mu_o \omega_o c}{2n} p^{(2)} \exp(-j\Delta kz) \tag{5.24}$$

In the previous equation, the phase-matching part is $\Delta k = k_{2\omega} - 2k_\omega$ for SHG procedure. In SHG, under $A_\omega = $ constant, the integration over the sample length L of the previous equation, by considering initial condition like $A_{2\omega}(z=0) = 0$, will have the following equation:

$$A_{2\omega}(z=L) = -j\alpha_{2\omega}\chi^{(2)}A_\omega A_\omega L \exp\left(-j\frac{\Delta kL}{2}\right)\text{sinc}\left(\frac{\Delta kL}{2}\right) \tag{5.25}$$

In Equation (5.25), $\alpha_{2\omega} = \frac{\omega_o}{nc}$ is constant. The calculated SHG intensity is given as,

$$I_{SHG} = |A_{2\omega}|^2 = \alpha_{2\omega}^2 |\chi^{(2)}|^2 I_\omega^2 L^2 \text{sinc}^2\left(\frac{\Delta kL}{2}\right) \tag{5.26}$$

As regard to SHG method, Equation (5.26) freely deduces some important facts. Here, SHG intensity and the square of the input intensity I_ω are proportional. And just for ordered non-centrosymmetric frameworks, the $\chi^{(2)}$ is nonzero. The dependence of the SHG intensity on the susceptibility $\chi^{(2)}$ is quadratic, i.e., its dependence is N^2 where N represents the number density (Plotnikov et al. 2006). Moreover, the SHG intensity and the square of the intensity length L are also proportional to one another, given that the synchronization of phase is actually fulfilled ($\Delta k = 0$). But when this is not equal to zero then the dependency is represented through $\text{sinc}^2\left(\frac{\Delta kL}{2}\right)$ term.

When $\Delta k = 0$, the SHG is ideally phase matched in interfaces and uniaxial crystals provide the complete SHG signal in forward direction. Besides this, due to scattering and inherent randomness in the genetic tissues, the SHG method never totally fulfills the phase coordination (Figure 5.1(d)). This outcome results in irregular dispersion of Δk values inside an anisotropic field comprising both single fibril and a gathering of little fibrils. Subsequently, the discharge has a circulation in forward and in reverse ways (Matteini et al. 2009).

5.4.2   Third Harmonic Generation (THG)

There is also another nonlinear multiphoton imaging method that is called THG, which discovers contrast in nearby differences in third-order susceptibility $\chi^{(3)}$ or in nonlinear RI (Barad et al. 1997; Squier et al. 1998). Such alterations normally happen at primary interfaces (Tsang 1995), for

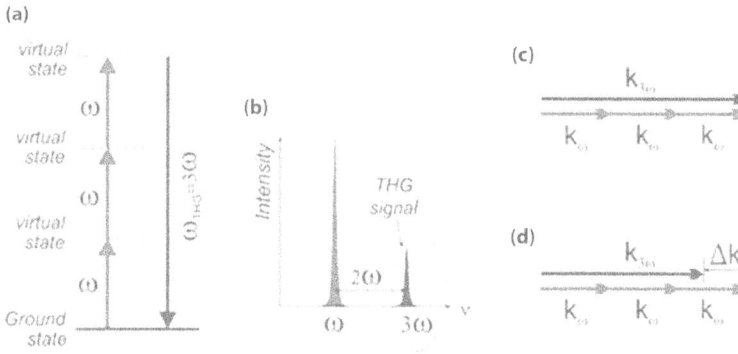

FIGURE 5.2 THG procedure: (a) energy level diagram, (b) spectral positions of input and output fields, (c) perfect phase matching, and (d) normal wave-vector misalignment that is recompensed through nonlinear refractive index varying at structural interfaces (Kumar, Coluccelli, and Polli 2018).

instance, at water-protein and water-lipid boundaries in lipid dewdrops, collagen packets, extra- and intracellular films, and furthermore at interfaces of inorganic designs.

In Figure 5.2(a), when a beam of laser strikes the material, three photons that are at laser optical frequency ω are interceded through a simulated state of the structure and up changed over into a single photon that having optical frequency 3ω that is referred to as THG cycle. It is also known as a four-wave mixing (FWM) coherent scattering technique. The THG also begins from the induced polarization instead of any actual absorption, similar to SHG. Third-order induced polarization producing THG signal is given by the following equation:

$$P_{THG}^{(3)} = 2\varepsilon_0 \chi^{(3)} \left(-3\omega; \omega, \omega\right) \cdot A_\omega A_\omega A_\omega \exp\left(j3\omega t\right) \tag{5.27}$$

The THG intensity can be represented as

$$I_{THG} = \left|A_{3\omega}\right|^2 = \alpha_{3\omega}^2 \left|\chi^{(3)}\right|^2 I_\omega^3 L^2 \sin c^2 \left(\frac{\Delta kL}{2}\right) \tag{5.28}$$

In Equation (5.28), the phase-matching term is $\Delta k = k_{3\omega} - 3k_\omega$ and $\alpha_{3\omega} = \frac{\omega_0}{nc}$ is a constant for the THG procedure. The distinctive characteristics of THG signal are shown by Equation (5.28). Here, THG intensity and the third power of the input intensity I_ω are proportional to one another. The dependence of the THG intensity on the susceptibility $\chi^{(3)}$ is quadratic, for example, it is corresponding to N^2. Moreover, the THG intensity and the square of the intensity length L are also proportional to one another, given that the synchronization of phase is actually fulfilled $(\Delta k = 0)$. But when this is not equal to zero, then the dependency is represented through $\sin c^2 \left(\frac{\Delta kL}{2}\right)$ term.

THG appears to be more adaptable than SHG as THG isn't restricted by a specified asymmetry or structures, in spite of the fact that SHG requires some non-centrosymmetric structure (Sun et al. 2000). In any way, the THG signal is not produced by the homogenously scattered medium, because in this case the term $\chi^{(3)} \left(-3\omega; \omega, \omega, \omega\right)$ disappears. As an alternative, the term $\chi^{(3)} \left(-3\omega; \omega, \omega, \omega\right)$ is nonzero that is provided through the optical inhomogeneities of scopes of the demand of laser focal volume and in addition it is used to produce the THG. In this manner, THG signal strength relies upon the amount around a specific specimen structure size (Yelin et al. 1999).

5.5 SECOND-ORDER NONLINEARITY EFFECT

In non-centrosymmetric framework, the second-order nonlinearity is the most reduced order of nonlinear interaction. The third wave is created by mixing two input waves. The equivalent amplitudes of two optical fields of frequencies ω_1 and ω_2 are connecting with each other in the medium, here the greater of both frequencies is ω_1 (Kleinman 1962a). The resulting optical field amplitude is given as

$$E(t) = \frac{1}{2}E(\omega_1)e^{-i\omega_1 t} + \frac{1}{2}E(\omega_2)e^{-i\omega_2 t} + \text{c.c.} \tag{5.29}$$

The second-order polarization in the explicit form is given as

$$\begin{aligned}
\mathbf{P}^{(2)}(t) = &\frac{1}{4}\chi^{(2)}\left(2\omega_1; \omega_1, \omega_1\right)E^2(\omega_1)e^{-i2\omega_1 t} + \frac{1}{4}\chi^{(2)}\left(2\omega_2; \omega_2, \omega_2\right)E^2(\omega_2)e^{-i2\omega_2 t} \\
&+ \frac{1}{4}\left[\chi^{(2)}\left(\omega_1 + \omega_2; \omega_1, \omega_2\right) + \chi^{(2)}\left(\omega_2 + \omega_1; \omega_2, \omega_1\right)\right]E(\omega_1)E(\omega_2)e^{-i(\omega_1+\omega_2)t} \\
&+ \frac{1}{4}\left[\chi^{(2)}\left(\omega_2 - \omega_1; \omega_2, -\omega_1\right) + \chi^{(2)}\left(-\omega_1 + \omega_2; -\omega_1, \omega_2\right)\right]E(\omega_2)E_{(\omega_1)}^*\left\langle e^{-i\omega_1 t} | e^{-i(\omega_2-\omega_1)t}\right\rangle \\
&+ \text{c.c.} + \frac{1}{4}\left[\chi^{(2)}\left(0; \omega_1, -\omega_1\right) + \chi^{(2)}\left(0; -\omega_1, \omega_1\right)\right]\left|E(\omega_1)\right|^2 \\
&+ \frac{1}{4}\left[\chi^{(2)}\left(0; \omega_2, -\omega_2\right) + \chi^{(2)}\left(0; -\omega_2, \omega_2\right)\right]\left|E(\omega_2)\right|^2
\end{aligned} \tag{5.30}$$

As demonstrated in Equation (5.30), the 16 susceptibility tensors are utilized to explain the second-order nonlinear polarization. The SHG is represented by the first two terms in Equation (5.30), where two photons having similar frequency (ω_1 in first term and ω_2 in second term) work together inside a nonlinear medium to create a new photon with double the energy than incident photons (Kleinman 1962b). The SFG is represented by the third and fourth terms. The third photon is created by the extinction of the two photons with dissimilar frequencies. To create the DFG, the two photons that have dissimilar frequencies get destroyed in the formation approach as depicted in the fifth and sixth terms of Equation (5.30).

In the fifth and sixth terms, the bracket symbolization $\left(\left\langle e^{-i\omega_1 t} | e^{-i(\omega_2-\omega_1)t}\right\rangle\right)$ is utilized to represent the creation of frequency constituents (Rice et al. 1994). The new difference frequency term, $e^{-i(\omega_2-\omega_1)t}$, is represented in the part of ket, and the residual term of frequency, $e^{-i\omega_1 t}$, in the DFG is represented by the component of the bra in the sixth term of Equation (5.30) (Giordmaine and Miller 1965).

The finest example of DFG is the optical parametric amplification (OPA). The last term in Equation (5.30) resembles OR. In 1962, it was initially reported in the non-centrosymmetric medium that it gave the semi-static polarization (He and Liu 1999). By taking all susceptibility terms together with nonlinear susceptibility tensors, the whole number of components is 432{[54 + 54 (c.c.) (SHG)] + [54 + 54 (c.c.) (SFG)] + [[54 + 54 (c.c.) (DFG)] + 108 (OR)]}(Auston, Glass, and Ballman 1972).

5.6 RESONANT THIRD-ORDER NONLINEAR EFFECTS

When light interacts with matter, numerous phenomena are considered such as light scattering, luminescence, refraction, and absorption. Optical properties of a material become the function of intensity at high values of intensity. Such interactions have been studied in the field of NLO (Christodoulides et al. 2010). There is a four-wave combining process in third-order NLO where three waves combine and create the fourth wave. Three waves of frequencies, ω_1, ω_2, and ω_3, in third-order nonlinear process have an optical amplitude of

$$E(t) = \frac{1}{3}E(\omega_1)\text{fe}^{-i\omega_1 t} + \frac{1}{3}E(\omega_2)e^{-i\omega_2 t} + \frac{1}{3}E(\omega_3)e^{-i\omega_3 t}\text{c.c.} \tag{5.31}$$

SC optical devices that utilize NLO have increased with years as the response time of optical nonlinearities can diverge from fs to milliseconds in SCs (Rao 2016). Third-order nonlinear effects in SCs provide absorption coefficient and RI that depends upon intensity. Resonant nonlinear effects are produced by photo-generated transporters and deal with energies of photon near to the absorption edge in an SC. In resonant effects, real photo-generated carriers are created that provide optical nonlinearities. A number of mechanisms in SCs take place when photon energies are near to bandgap resonance such as band filling, exciton shifting, and lightening on low-dimensional SCs. The effect of band filling is shown in Figure 5.1. The absorption coefficient depends upon F_c and F_v that are Fermi functions. When light falls on SCs as a result of photon absorption, electron-hole pairs are generated and the Fermi functions will change. Electrons in conduction band exchange energy with lattice quickly and achieve quasi-thermal equilibrium in less than 300 fs (Al-Hemyari, Ironside, and Aitchison 1992). Holes in the valance band do a similar process. It takes a long time to achieve thermal equilibrium and electrons to come to the valance band from the conduction band because of inter-band relaxation that takes place about 10 ns. Fermi functions are:

$$F_v = \frac{1}{e^{\left(\frac{(E_v - E_{Fv})}{KT}\right)} - 1} \tag{5.32}$$

and

$$F_c = \frac{1}{e^{\left(\frac{(E_v - E_{Fc})}{KT}\right)} - 1} \tag{5.33}$$

Net absorption coefficient becomes

$$\alpha(E_t) = \frac{\pi e^2 h |Pcv|^2}{\varepsilon_0 \, cm_0^2 E_t} \int_{E_g}^{\infty} N_{cv}(E_t) \delta(E_t - h_\omega)[1 - F_v]F_c - (1 - F_c)F_v]dE_t \tag{5.34}$$

This becomes

$$\alpha(E_t) = \frac{\pi e^2 h |Pcv|^2}{\varepsilon_0 \, cm_0^2 E_t} \int_{E_g}^{\infty} N_{cv}(E_t) \delta(E_t - h_\omega)[F_c - F_v]dE_t \tag{5.35}$$

In terms of Fermi's Golden rule with a change in Fermi function, the density of states is changed but coupling between states remained constant so absorption coefficient expression changed as represented by Equation (5.35). At photon energies of higher intensities near to or above bandgap, there is a transferring of bandgap energy to a higher energy that is called blue shift. Thus, the absorption near the bandgap decreases. In a good quality SC, inter-band relaxation time that is of few nanoseconds is used to find recovery time of absorption (Christodoulides et al. 2010). When photon energies are lesser than bandgap resonance then variation in RI will be negative. The extension of electric susceptibility cannot be utilized where real photo carriers are generated during resonance nonlinearity so there is a need to modify the electric susceptibility by integrating it over the region of proper space and time. There are some effects of a limited lifetime, including the one that no longer has the dependency upon the pulse intensity as there is equilibrium between the rate of formation and the recovery of carriers. For CW (continuous waveform) excitation nonlinear effects have direct relation with intensity but for pulse excitation, these effects are found out by the relationship between recovery time and pulse width (Ironside 1993). Several photo-generated carriers, n_g, can be found out by the following relation:

$$n_g \propto \int_0^{\tau_r} f(t)dt \qquad (5.36)$$

where τ_r represents the recovery time of carriers and $f(t)$ represents pulse shape. When the pulse breadth is shorter than recovery rate then the carriers have a direct relation with a total amount of photons in pulse and not with intensity. A high-intensity ultrashort pulse of fs contains no more photons; therefore, no more real photo carriers are generated (Ironside 1993). So, the effect of band-filling nonlinearity cannot be produced by ultrashort high-intensity pulse. Saturation and thermal effects also came into mind for resonant nonlinearity. Saturation arises when for a given photon, the intensity of the light is enhanced then absorption rate will also change; this is because either early states are all empty or final states all are completed. An increase in temperature will increase the RI. This thermally altered RI has an opposite sign to electronic nonlinearity. SC micro-crystallites show non-radiative combination centers of electrons and hole carriers by reducing inter-band relaxation time because of defects associated with them.

Because of nonradioactive defects, heat is produced that gives rise to unwanted thermal effects such as a replace in index of refraction. The refractive portion is calculated by the pump-probe interferometry approach. When results were plotted between induced RI and pump flounce at different values of recovery time, it was observed that at short times there was a negative effect in RI that was because of the effect of band filling, whereas the positive effect in RI was because of thermal effects. These two effects have the opposite signs. General performance was normal of resonant nonlinearity in SCs and has several limitations to its effectiveness.

5.7   NONRESONANT THIRD-ORDER NONLINEAR EFFECTS

In nonresonant effects, photon energies are under the basic absorption edge. Nonresonant optical effects indicate intensity-dependent absorption and IDRI at photon energies below bandgap energy [6]. Nonresonant nonlinearity has a rapid response than resonant nonlinearity but is smaller than resonant nonlinearity. Nonresonant refractive nonlinearity is related to two-photon absorption. Third-order nonlinear susceptibility $\chi^{(3)}$ that provides nonlinearity in RI can be explained by the following equation:

$$n = n_0 + n_2 I \qquad (5.37)$$

Intensity-dependent part of the RI is explained by a coefficient n_2. The nonlinearity in RI is related to a real part of $\chi^{(3)}$ and the absorption coefficient of two photons is related to the imaginary part of $\chi^{(3)}$. Both real and imaginary parts of $\chi^{(3)}$ are connected through Kramers-Kronig relations (Krowne 2019), where the nonlinearity in RI is measured by the frequency dependence of two-photon absorption coefficient. These have been applied to different SCs and insulators like silica. In SCs for one-photon absorption, not only photon energy should be greater than bandgap energy but also it should be in the range of the following:

$$\frac{E_g}{2} < h\omega < E_g \qquad (5.38)$$

If the light has more intensity, then this is called two-photon absorption. In this absorption process, both photons are either emitted or absorbed during the transition in the material. It is a process where one photon is responsible for transition in the intermediate virtual level and the other photon forms the transition from the virtual level to the final level. A virtual level in an atom is an intermediate energy level that has momentary lifetime with respect to lifetime of other levels existing in the substance. Two-photon absorption can also occur because of the high EF of light that changes the optical properties of the material. Two-photon absorption is considered intensity-dependent absorption by the following equation:

$$\frac{dI}{dz} = -\left(\alpha I + \beta_2 I\right) \tag{5.39}$$

In the previous equation, z, β_2, and α denote propagation direction, two-photon absorption coefficient, and linear coefficient of absorption, respectively, and I is the intensity of light. β_2 can be represented as

$$\beta_2 = \left(\frac{2\hbar\omega_p}{I^2}\right) M_{12} \tag{5.40}$$

In two-photon absorption coefficient Equation (5.40), ω_p denotes operating frequency and M_{12} is transition rate from valance to conduction band [8]. Transition rate works from second-order perturbation theory, that is,

$$M_{12} = \frac{2\pi}{h} \left(\frac{4\pi^2 e^2 I^2}{n_p c^2 m_0^4 \omega_p^4}\right) \sum \int_{-\infty}^{\infty} \frac{\left|\langle C|\mathbf{e.p}|i\rangle\langle i|\mathbf{e.p}|V\rangle\right|^2}{E_i - E_V - h\omega_p} * \delta\left(E_C - E_V - 2h\omega_p\right) \frac{dk}{(2\pi)^3} \tag{5.41}$$

where C, c, i, and V denote conduction band, speed of light, intermediate band, and valance band, respectively, h is the plank constant, n_p is the RI, and the term $\sum \int_{-\infty}^{\infty} \left|\frac{\langle C|\mathbf{e.p}|i\rangle\langle i|\mathbf{e.p}|V\rangle}{E_i - E_V - h\omega_p}\right|^2$ is called photon coupling coefficient. Equation (5.41) is also the second-order form of Fermi's Golden rule that states that transition rates have a direct relation with coupling between states and density of states. It shows that coupling depends upon intensity for two-photon absorption. Delta function in Equation (5.41) shows that only those states couple that satisfy the energy conservation rule. The density of states and two-photon coupling between states can be found out by the band explanation of a substance and from Kane's theory SCs along with second-order perturbation theory, respectively (Klingshirn 1990). After considering the density of states and coupling between two photons using some approximations, two-photon absorption coefficient expression becomes

$$\beta_2 = 3.1 \times 10^3 \frac{\sqrt{E_p}}{n^2 E_g^3} F\left(\frac{2h\omega_p}{E_g}\right) \tag{5.42}$$

In the above equation, E_p and E_g denote Kane momentum and energy, respectively, that are almost constant for a variety of SCs. If both E_g and E_p are in eV then constant 3.1×10^3 provides β_2 coefficient in $^{cm}/_{GW}$. Function F(x) can be found out by the band structure of SCs like for parabolic energy bands this function is given by:

$$F(x) = \frac{(x-1)^{\frac{3}{2}}}{x^5} \tag{5.43}$$

The previous equation can be applied to a variety of SCs. At first, it was derived only for III–V SCs but because of its wide-ranging applicability, it also showed good results for II–VI SCs and insulators. An equation that shows the magnitude and dispersion of n_2 is

$$n_2(esu) = \frac{K' G_2\left(\frac{h\omega}{E_g}\right)}{n_0 E_g^4} \tag{5.44}$$

In the previous equation, $K' = 3.4 \times 10^{-8}$ and G_2 is given by

$$G_2(x) = \frac{-2 + 6x - 3x^2 - x^3 - \frac{3}{4}x^4 - \frac{3}{4}x^5 + 2(1-2x)^{\frac{3}{2}}\theta(1-2x)}{64x^6} \tag{5.45}$$

where θ shows a step function. Both Equations (5.44) and (5.45) are used to find out nonlinear refraction for a variety of materials. Above all, relations show that two-photon absorption and non-resonant nonlinearity are related to one another through the Kramers-Kronig relations for a variety of solids (McGill, Torres, and Gebhardt 2012). So, in SCs, third-order nonlinearity is accurate for both two-photon absorption coefficient β_2 and RI nonlinear part, n_2. The influence of the EF on SCs is instantaneous and linear absorption is kept low so that devices that show nonresonant nonlinearity must have a high transmission of light.

5.8 NONLINEARITIES IN SCs

In NLO, all-optical properties of a material are dependent upon the intensity of incident light inversely. On this basis, there are two groups of optical nonlinearities. In the first group, the system reacts with amplitudes of electric and magnetic fields E_i. This condition is satisfied only for virtual excitations that arise because of uncertainty in energy and time. For this condition, susceptibility expands in terms of power series of EFs such as

$$\frac{1}{\varepsilon_0}P_i = \chi_{ij}^{(1)}E_j + \chi_{ijk}^{(2)}E_jE_k + \chi_{ijkl}^{(3)}E_jE_kE_l + \ldots \tag{5.46}$$

On right side of the previous equation, the terms indicated as (1), (2), and (3) represent linear optics, second harmonic, and wave mixing, respectively. All processes are coherent as there exists a phase relationship between **P** and various values of **E**. In this case RI in terms of intensities can be written as Equation (5.37). All the previous approaches fail when we take real excitations into account that have finite lifetime τ. In this case, susceptibility is dependent upon the density of exciting species $n_p(t)$, which can be phonons, electron-hole pairs, or excitons. Thus, we achieve the following equation of the form:

$$\chi = \chi(\omega, n_P), \quad n_p = \int_{-\infty}^{1}G(t')e^{\frac{-t'}{\tau}}dt' \tag{5.47}$$

where $G(t)$ represents the generation rate that depends upon the intensity of incident light for one- and two-photon excitations. This is explained as

$$G(t') = \alpha(\hbar\omega_{exc})I_{exc}(t') + \beta(\hbar\omega_{exc})I_{exc}^2(t') \tag{5.48}$$

All coefficients in Equation (5.48) like α and β depend upon $n_P(t)$ that give rise to coupled differential equations. In this case, all processes explained by Equations (5.47) and (5.48) are called incoherent and lose their phase memory during their lifetime because real excited particles undergo the scattering process. Nonlinearity in SCs can also be classified by considering another aspect. If in electronic system nonlinearity exists because of real and virtual optical excitations, then these denote photoelectronic optical nonlinearities. If these are because of sample heating due to absorbed light during non-radiative recombination, then these denote photothermal optical nonlinearities. At last, if we limit the frequency of light field to zero, then we have the effect of DC voltage at which bridge to optoelectronics opens. We will consider here only SCs from groups of III–V and III–VI that have direct bandgap. It is necessary to know what happens with amplifying excitation intensity due to real and virtual electron-hole pairs or due to photothermal nonlinearities. In all SCs, by increasing excitation density, three regions are observed. In linear regions, optical properties of the material are determined by individual exciton. Carriers move freely along the whole crystal or are bound at low temperatures. The density of carriers is very low in this region that they do not interact with each other in their lifetime. In middle region, excitons start to interact with each other (Yan et al. 1990). These interactions can be either elastic or inelastic scattering between excitons or among free particles and excitons. This scattering increases the exciton resonances via the reduction in relaxation time and thus generating some bands of induced absorption. Biexcitons also give rise to optical nonlinearities that are states of two electron-hole pairs. In the third region, electron-hole pair's density is so high that their distances become comparable to the Bohr radius. In this regime, a combined phase is considered that is called EHP. There are some examples of optical nonlinearities of SCs such as photoelectronic optical nonlinearity and photothermal optical nonlinearity.

5.8.1 Photothermal Nonlinearities

As explained in Equation (5.47), optical nonlinearities can also be a result of the increasing number of phonons by absorption of light. When photons of light are sent to SCs around absorption edge, these will interact with the system as a result producing virtual and real electron-hole pair states. Recombination is non-radiative in most SCs; therefore, most of absorbed light is changed into heat. If heat is enough to enhance temperature of lattice T_L, then optical properties of the material will be affected. Absorption coefficient, α, in all SCs follows the Urbach-Martienssen rule that is given as

$$\alpha(\hbar\omega, T_L) = \alpha_0 \exp[-\sigma(T_L)(E_0 - \hbar\omega)(k_B T_L)^{-1}] \text{ for } \hbar\omega < E_{ex} \qquad (5.49)$$

In Equation (5.49), α_0, E_0, and $\sigma(T_L)$ denote material parameters. Phonons give rise to thermal distribution that is represented by lattice temperature T_L that is because of strong anharmonicity, so there is need to replace n_P with an increase of lattice temperature in Equation (5.47) (Schöll and Scholl 2001). These photothermal nonlinearities will increase the magnitude of α and n as explained in Equation (5.49) on a low-energy absorption edge. The nonlinear dynamic effects were observed when an induced absorptive bistable element was carried into hybrid ring resonator at τ_R that was the round-trip time and was larger enough than the thermal relaxation time τ_t. There was an Ikeda resonator in the system that consisted of a nonlinear dispersive element. By keeping input intensity constant, a variety of self-oscillations were viewed. Self-oscillations were produced as intensity was raised step by step in resonator for various τ_R until the system switched to an absorbing state. At absorbing state intensity of intracavity was absorbed, and the system reverts back to the transparent state and entire process begins again. By increasing incident intensity, various modes of oscillation were observed (Schmitt-Rink, Chemla, and Miller 1985). These modes were made according to the Farey tree. Stability of variety of modes as a function of I_0 obeyed a devil's set of steps. Recently, irregular oscillations were seen in optical ring resonators consisting of three absorptive bistable elements.

5.8.2 Photoelectric Optical Nonlinearities (PEONL)

According to literature, optical nonlinearities have been related with the elastic and inelastic scatterings (Klingshirn and Haug 1981). In photoelectric nonlinearities, the biexciton states transitions are involved (Hönerlage et al. 1985). Recently, the biexcitons were found in the II–VI quantum well, as per this, giving the increment of the biexciton-binding energy anticipated from the suitable increment of exciton-binding energy, known in quasi-2D frameworks (Gunshor and Kolodziejski 1988; Klingshirn 1990). The types of nonlinearities that result from strong optical excitations of the SC materials are named the PEONL. In PEONL nonlinearities, the main concern remains with the band-to-band transition, mostly materials belonging to two groups II–VI and III–V, direct bandgap of the SC materials. The PEONL are mostly found in quasi-2D or three-dimensional (3D) SC materials. In the low excitation section, the optically excited electron-hole pairs, the density is minimal due to which they don't interact with each other. The formed excitons are either certain to some defect or free at particular low temperature. These exciton states have been utilized for the determination of the optical properties at the absorption edge's vicinity. As increment in temperature occurs, as a result of phonon interaction with excitons, ionization occurred. But still at band-edge, the Coulomb interaction as well as oscillator strength influence exists (Hönerlage et al. 1985; Klingshirn and Haug 1981).

In the intermediate density section, the process of the interactions among excitons starts. In this, the collisions may be elastic or inelastic or combined to build a new excitonic state named biexciton. These scattering processes of biexciton state with interaction of photon give rise to the PEONL such as excitation-induced absorption, or new emission band arrival, or exciton resonance's collision broadening (Broser and Gutowski 1988; Hönerlage et al. 1985).

In high-density section, the average distance of electron-hole pair becomes comparable to the excitonic Bohr radius. So, the EHP phase is formed due to breakdown of the quasiparticles. Due to EHP transitions, significant change in the optical properties occurred (Klingshirn et al. 1989).

5.9   LINEAR ELECTRO-OPTIC (LEO) EFFECTS

In this case, a material is exposed to an EF that is static along with a single wavelength light having frequency w. In such a way that static field plays a role in oscillating at a frequency which is below as compared to optical frequency. Consequently, these results also lie under the scope of Equation (5.8), which leads to a modified optical phenomenon of the induced EF and also named the electro-optical effects. At zero frequency when EF is applied, no changes appear in output frequency after modification.

The LEO effect that is also named Pockels effect comes from the non-centrosymmetric material's quadratic susceptibility $\chi^{(2)}(-\omega;\omega,0)$. The linear electric susceptibility $\chi(w)$ has dependency on the $\mathbf{P}(\omega)$ as well as $\mathbf{E}(\omega)$ from Equations (5.4) and (5.5). Because susceptibility tensor, $\chi(w)$, varies only by the constant unit tensor from the relative permittivity tensor, $\varepsilon_r(\omega)$, the identical quantity also yields $\varepsilon_r(\omega)$ linear EF dependency. The refractive indices are determined by relative permittivity, and the same amount causes their linear EF dependency. Practically, EF dependency on optical indicatrix ε^{-1} is encountered. In the case of the principal component, ε_r corresponds to ε_i and then light that propagates is parallel and polarized parallel to the principal axes i and j, respectively. Then corresponding RI is $n_j = \varepsilon_j^{\frac{1}{2}}$. The equation $v \cdot \varepsilon_r^{-1} \cdot v = 1$ in terms of the principal axes is written as follows:

$$\left(\frac{\mathbf{v}_1^2}{\mathbf{n}_1^2}\right)+\left(\frac{\mathbf{v}_2^2}{\mathbf{n}_2^2}\right)+\left(\frac{\mathbf{v}_3^2}{\mathbf{n}_3^2}\right) \tag{5.50}$$

Apart from optical absorptions, $n_i \geq 1$, Equation (5.50) signifies an ellipsoid as in the present case v vector and principal axes are parallel to each other while principal refractive indices are represented by the semiaxes length. If the principal axis is 1 and v is parallel to it, then corresponding values are $v_2 = 0 = v_3$ and $v_1 = \pm n_1$. Moreover, the RI is obtained by corresponding vector length v from the surface origin of ellipsoid (Landau and Lifshitz 1984; Rohleder and Munn 1992).

For the determination of dependency of the EF ε_r^{-1}, considering tensor as Q, we need to consider a matrix M and solve it for obtaining the results:

$$\frac{d\mathbf{M}^{-1}}{dx} = -\mathbf{M}^{-1} \cdot \frac{d\mathbf{M}}{dx} \cdot \mathbf{M}^{-1} \tag{5.51}$$

And differentiating and rearranging the Pockels coefficient that represents the relation with the linear field are described as follows:

$$r_{\alpha\beta\gamma} = \partial Q_{\alpha\beta} / \partial E_\gamma \tag{5.52}$$

$$r_{\alpha\beta\gamma} = -Q_{\alpha\lambda}\, \chi_{\lambda\mu\gamma}\left(-\omega;\omega,0\right) Q_{\mu\beta} \tag{5.53}$$

As in the case of the principal axes, $Q_{\alpha\beta} = \chi_{\alpha\beta}/n_\alpha^2$, the previous equation reduces as follows:

$$r_{\alpha\beta} = \chi_{\alpha\beta\gamma}\left(-\omega;\omega,0\right) n_\alpha^2 n_\beta^2 \tag{5.54}$$

Here, the repeated indices are α and β that cannot be added up. Due to this reason, the same indices after interchanging the value of LEO effect coefficient's frequency remain unaffected. The possible combination of the Voigt notation is six, such as $r_{123} \rightarrow r_{63}$. The Pockels effect can be portrayed by the utilization of the coefficient $r_{\alpha\beta\gamma}$, which gives us dependence on linear field of indicatrix, while the Kerr effect can be elaborated via the utilization of coefficient $P_{\alpha\beta\gamma\delta}$, which gives us dependence on the quadratic field. Similar to the quadratic effects, the LEO effect is missing in the centrosymmetric materials. One would then be able to respect the Kerr impact in such materials as an electric-field-incited LEO impact, whereby EF initially contorts materials as well as annihilates its centrosymmetrics and afterward enrolls an LEO impact in distorted materials. Additionally, there is a variant of each $X^{(2)}$ wonder that is incited in a centrosymmetric material by means of applying of static EF via an appropriate $X^{(3)}$.

5.10 INTENSITY-DEPENDENT RI

The more general phenomenon is IDRI that is needed to study the interactions of waves having different frequencies. The AC Kerr impact relates to an adjustment of the direct optical conduct of medium at one frequency by the utilization of different frequencies of intense beam. What if the two frequencies are the same? There is no additional applied beam in this case: the phenomenon is caused by a single beam at a single frequency through the coefficient $\chi^{(3)}\left(-\omega;\omega,-\omega,\omega\right)$. This relates with RI and is altered via using DC part of $E(\omega)^2$.

Nonlinear RI expressed as

$$n = n_0 + n_2 I \tag{5.55}$$

where n_0, n_2, and I represent the refractive indices of low intensity, nonlinear coefficient and intensity, respectively. The RI in terms of amplitude $|E|$ of the EF can be articulated as follows:

$$n = n_0 + n_2' |E|^2 \tag{5.56}$$

For the weak nonlinear case, the intensity is described as

$$I = \frac{1}{2}\varepsilon_0 n_0 c |E|^2 \tag{5.57}$$

Two nonlinear coefficients are closely related with one another by the following equation:

$$n_2' = \frac{1}{2}\varepsilon_0 n_0 c n_2 \tag{5.58}$$

where n_2 and n_2' are two coefficients in which one is convenient with respect to experiment while the other with respect to $\chi^{(3)}$. In a simplified form,

$$\chi = \chi^{(1)} + \chi^{(3)} |E|^2 \tag{5.59}$$

where $n^2 = 1 + \chi$ and $n_0^2 = 1 + \chi^{(1)}$, so the value of the

$$n_2' = \chi^{(3)}/2n_0 \tag{5.60}$$

From the previous equation, two effects are arising from IDRI. The first effect is that the optical device's component such as the waveguide is deliberated in such a way that the n_0 have quantitatively and qualitatively diverse properties at diverse values of intensities, whereas the value of the n is different in comparison to the value of n_0. For instance, optical cavity depends upon the value of RI and consequently it behaves like a light beam's switch. Underneath certain circumstances, optical cavity likewise exhibits bistability. For sufficiently high intensity, the transmission shifts from a low-to-high value and continues at this level until the intensity is lowered by a limited amount. Due to this type of hysteresis, nonlinear cavity also behaves like an optical memory element.

The RI's intensity dependency can be deliberated by using the photorefractive effect. In this process, the value of the RI depends upon the incident light and is related to the incident light. The RI reaches its original value once the light beam is detached. This type of effect comes from electronic as well as the nuclear arrangement distortion of the materials. Moreover, the photorefractive effect is also beneficial for the migration as well as separation of the charges under the influence of light. These charges are trapped and as a result generate high EFs in their respective vicinities. As a result of EFs, the value of RI changes due to photorefractive effect.

5.11 NONLINEAR REFRACTION AND ABSORPTION

Nonlinear absorption or two-photon absorption (2PA) is an instantaneous absorption of two equal photons that have different frequencies to stimulate a molecule from ground to excited state and this absorption is called nondegenerate two-photon absorption. As this absorption directly depends on instantaneous absorption of two photons, the possibility of 2PA is proportionate to square of intensity of light; hence this process is non-optical. The energy difference between states (excited and ground) of molecule is equivalent or less than the sum of photon energies absorbed by two photons (Yin et al. 2000).

The absorption cross section of 2PA being a third-order process is normally many orders smaller as compared to the one-photon absorption.

Two-photon absorption is an optical nonlinear (ONL) operation. The magnitude of 2PA in a particular molecule is connected to the third-order imaginary portion of the nonlinear susceptibility. Therefore, the transition rules for 2PA are different from one-photon absorption (Gopalakrishnan and Narendar) that depends on first-order susceptibility (Chemla et al. 1984).

The intensity decay due to OPA is defined by Beer's law as

$$I(x) = I_0 e^{-\alpha x} \tag{5.61}$$

where x is distance that is traveled by light through materials, $I(x)$ is the intensity of light when it travels distance x, and I_0 is the intensity of light when it enters in a material and α is OPA coefficient of the material. In 2PA, the intensity of light (I) versus distance (x) changes for the incident wave of radiations:

$$I(x) = \frac{I_0}{1 + \beta x I_0} \tag{5.62}$$

This Equation (5.62) is for 2PA, having intensity of light as the function of cross section x, initial intensity of light I_0, and concentration c. The coefficient of absorption α is now 2PA coefficient β (Ganeev 2005).

Several methods can be used for the measurement of 2PA. Some of them are Z-scan, nonlinear transmission, two-photon excited fluorescence and self-diffraction. 2PA, a third-order nonlinear absorption's process, is most effective at very high intensity, so pulsed lasers are more commonly used.

When effective beam of laser travels through a sample, its EF may cause a variation in material's RI that has direct relation with beam's strength. This effect is called the Kerr effect. The maximum RI of the surface is the sum of the RI n_o, when there is no laser beam existing and the expression n_2I, where n_2 deliberates second-order nonlinear' s RI, while I represents beam intensity:

$$N = n_o + n_2I \tag{5.63}$$

The shift in RI can be negative or positive. n_2 values are typically low (around 3×10^{-16} cm^2 W^{-1} for silica), high beam intensities are needed to have a significant impact (Christodoulides et al. 2010).

All of the approximately equal susceptibility tensor components are zero in inversion summery media. For a monochromatic pulse traveling in the z-axis with frequency ω, phase $\varnothing$, wave vector $k = n_o\omega/c$, and amplitude A,

$$E = A\,e^{i\varnothing}e^{i(kz-\omega t)} \tag{5.64}$$

Irradiance is given as

$$I = \frac{n_o c\varepsilon_o}{2}A^2 \tag{5.65}$$

where ε_o denotes free-space permittivity. Using the thin material and slowly changing envelope estimations, we can derive coefficients for the transition of irradiance and maximum phase within the medium by applying Maxwell's equations to Equation (5.64) (Boyd 2008).

$$\frac{dI}{dz} = -\alpha I \tag{5.66}$$

$$\frac{d\Delta\varnothing}{dz} = nk_o \tag{5.67}$$

where n is RI and α is absorption constant given as

$$\alpha = \alpha_o + \alpha_{NL} \tag{5.68}$$

$$n = n_o + n_{NL} \tag{5.69}$$

where n_o and n_{NL} are linear and nonlinear refractive indices, respectively, and α_o and α_{NL} are linear and nonlinear optical absorptions, respectively. Only nonlinear effects cause a phase shift:

$$\frac{d(\Delta\varnothing)}{dz} = n_{NL}k_o \tag{5.70}$$

n_{NL} and α_{NL} depend on input EF. The nonlinear refraction and absorption are as follows:

$$\alpha_{NL} = \alpha_2 I \tag{5.71}$$

$$n_{NL} = n_2 I \tag{5.72}$$

Equation (5.71) explains two-phonon absorption and Equation (5.72) represents nonlinear Kerr optical effect (Peceli 2013).

Different techniques have been employed to study these effects. The Z-scan technique has been used to study two optical absorption, charge carrier's absorption, nonlinear refraction, and refraction emission spectra for different SCs, for example, GaAs, InP, and InAsP. Although the form of the two-photon maximum absorption follows the theoretical estimate, the values calculated with the picosecond method are off two times. Theoretically, nonlinear refraction and charge carrier nonlinearities are all in strong agreement. With picosecond Z-scan observations, the actual value of third-order nonlinear effects in GaAs is also confirmed. Because of the fs laser's broad frequency range, the three-photon spectrum of absorption of GaAs was also studied with picosecond Z-scan. The spectral form agrees well with theory, but the measurements of several photon absorption cross sections are greater than predicted by theory (Capper, Willoughby, and Kasap 2020).

5.12   APPLICATIONS OF NONLINEAR OPTICAL ORGANIC SCs

In recent decades, the reliability of fiber optics has improved significantly, suggesting a tremendous growth of photonic communications systems. The findings of laser emissions have increased scientific performance in the area of uncommon optical effects in a range of materials. The data communication velocity determined by light defines an improvement in the effectiveness of devices used in a wide variety of applications, from networking to medication and in air force sectors, and also complete monitoring of the optical transmission level and without the requirement for an optical-electronic-optical transformation. Some of the samples are the displays of liquid crystals, organic materials for luminous display, organic light emitting diodes (OLEDs), products of piezoelectric plastics for headphones, and organic matter with special optoelectronic characteristics in the solid state for photonics applications (Tripathi et al. 2007).

ONL technologies are presumed to be essential components for potential photonic applications as these allow the control of a luminous beam and use another luminescent beam. The manipulation of the exceptional characteristics of organic molecules introduces them to unique and exciting technical areas (Babu et al. 2009). In fact, the key characteristics of the molecular structure specify the reaction (specific, rapid, and severe) at a macroscopic scale under any external stimulus action, such as luminescent radiation.

Carbon-based compounds are being extensively studied as they exhibit peculiar properties linked with both the π electrons in a conjugated configuration and groups introduced to the oxidative nucleus, resulting in intramolecular charge carrier properties important to a wide range of applications, particularly NLO, and providing a substitute with compounds that are inorganic. The production of thin films and crystals from oligomeric, molecular, and polymeric composites with conjugated process, generalized electron structures, and optical sensing nonlinear chromophore collections, displaying an organization of the fluorescent dyes sufficient for the presence of certain desirable molecular structure, is of particular attention in the development of ONL application (Stanculescu and Stanculescu 2015).

The valence electrons of the chromophores cause the formation of a broad delocalization of the electron cloud in the organic material, and functional groups are replaced with the oxidative nuclei that act as acceptors or electron contributors. Furthermore, to produce nonlinear optical effects, the organic molecules must be macroscopically packed in an asymmetrical manner. These situations may be observed in molecular crystals in bulk or thin film materials made of polymeric materials that have a high molecular order and chromophore group concentration (Ahlheim et al. 1996).

Special requirement for molecular crystals is the crystalline phase, while specific requirements for polymeric materials and monomers are the efficiency of nanostructures. The key features that make the use of organic molecular crystals suitable for the stated applications include a wide area of transparency, strong birefringence, high nonlinear coefficient, high laser light optical thresholds, and flexibility of the molecular structures that can be modified by molecular engineering to optimize the properties of interest. Since the huge exploration of organic materials exists at large in optical nonlinearities, the involvement of these materials has increased, opening up new directions for research and development in the field of photonics (Stanculescu et al. 2011).

5.12.1 Bulk Organic SCs

Aromatic derivatives benzil (Bz) and like *meta*-dinitrobenzene (*m*-DNB) are organic compounds that can be used to make bulk crystal substances with high optical bandgaps, large transparency domains, and large nonlinear coefficient. These two composites exhibit a significant optical nonlinear phenomenon of interest regardless of the difference in their chemical structures at the molecular scale (Stanculescu et al. 2005).

The significance in analyzing Bz and *m*-DNB bulk quartzes is demonstrated by the possibility of using these substances as a crystal mass matrix both for inorganic and organic guest guanidine in applications of ONL, which involves the production of organic composites, organic crystal development, and characterization. The adverse biaxial crystal *m*-DNB ($C_6H_4N_2O_4$), which corresponds to the Pb_{n21} crystal structure as well as the mm^2 point alignment group, crystallizes throughout as the orthorhombic system at ambient temperature, and has a pyramidal structure. The crystal structure has the succeeding dimensions: a = 13.20 Å, b = 13.97 Å, and c = 3.80 Å (Stanculescu, Ionita, and Stanculescu 2014).

m-DNB crystalline materials have transparency range of 0.4–2.5 μm. Bz ($C_6H_5COCOC_6H_5$) is a uniaxial crystal with the trigonal crystal structure $D_{34}(D_{36})$ and the $P_{3_12_1}$ line symmetry group. The hexagonal crystal lattice has the following components: a = 8.42 Å and b = 13.75 Å. It comprises three particles that are helically arranged and tightly bound round the 31 axes. The transparency array of Bz crystal structures is from ultraviolet to near infrared (Stanculescu, Ionita, and Stanculescu 2014).

The data from an X-ray diffraction of doped and pure Bz and *m*-DNB quartzes analyzed using the Pawley method and software TOPAS illustrated the impact of dopant upon their lattice constants. The experiment was carried out on slices that were cut right angles to the crystal's direction of propagation. The crystalline planes giving the greatest reflections were adjacent to the slice's surface, so the (003) plane gave the strongest reflection in Bz (Stanculescu 2007).

The crystal's growth direction is [001], and the (111), (311), and (002) planes offer the best *m*-DNB reflections, demonstrating the polycrystalline existence of the prepared samples with three regions of three different configurations. The hexagon shaped Bz crystal's lattice dimensions' values, a = b = 8.350 Å and c = 13.557 Å, agree with the calculated values a = b = 8.410 Å and c = 13.679 Å. The lattice constants determined for the *m*-DNB orthorhombic crystals, a = 13.246 Å, b = 14.029 Å, and c = 3.807 Å, also follow the reference values a = 13.290 Å, b = 14.070 Å, and c = 3.813 Å (Barvinschi, Stanculescu, and Stanculescu 2011).

The performance of the organic crystals, which is measured by the form of the liquid-solid interface as a result of the growing conditions, has a great effect on optical properties of *m*-DNB. Furthermore, homogeneity necessary for doped crystal is related to dopant implementation, which is directly impacted by the solid-liquid interface form. A partially convex liquid-solid interface is expected for high-quality crystals, but the interface deviations during growing process are expected due to variations in the heat capacity of the organic compound in the solid and liquid states (Meyers et al. 1994). Since the conductivity of heat coefficients in molten state is less than 1 for both Bz and *m*-DNB, a convex development interfaces with a reduced deviation from the planar production interface is expected in the case of Bz. This approves that high-quality Bz crystals are easier to produce than *m*-DNB crystals (Marder, Beratan, and Cheng 1991).

5.12.2 Electro-Optic Polymers

NLO SCs can be used for a wide range of applications, including amplification of optical signals supported by the electro-optic effect, microfabrication, imaging, detecting, and cancer therapy supported by multiphoton immersion, in which molecules take several photons at the same time. Electro-optic effect is that in which a material's RI varies due to an applied electromagnetic field. The research of second-order nonlinear optical effects within polarizable non-centrosymmetric materials started in earnest in the 1970s (Oudar and Chemla 1977).

Dipolar molecules constitute an important group of molecules that has gotten a great deal of attention in which polarization in one dimension within the molecules is simpler than those in the reverse direction. To describe the origins of hyper polarizability, β, in such structures and to help in the production of such molecules in several systems, Oudar and Chemla (1977) proposed that β could be well represented by a two-state model that could be used to direct the development of second-order NLO:

$$\beta\mu\left(\frac{\mu_{ge}^2\left(\mu_{ee}-\mu_{gg}\right)}{E_{ge}^2}\right) \tag{5.73}$$

where μ and E are the dipole matrix component and transition energy, correspondingly, between lowest energy state (g) and the very first highly authorized excited states (e) of charge transfer. In physics, the implementation of a $\left(\mu_{ee}-\mu_{gg}\right)$ term indicated that when electrons interact with the alternating EF of radiation, they tend to migrate in one direction over the other (Marder 2006).

To impose the electronic biases, compounds for second-order NLO applications were dependent on oxidative π-electron systems that were unsymmetrically end-capped by electron-contributing and receiving groups. It is known that 4-(N, N dimethyl amino)-4′-nitrostilbene (DANS) is a typical average NLO chromophore for which the two benzene chains and a double bond form a conjugated π-system that provides dipolar electron, and the dimethyl amino collection behaves as the donors and the nitro groups behave as the acceptors. The level of charge separation in the lowest energy state, or ground-state polarization, is primarily influenced by the chemical arrangement (e.g., the configuration of the π-conjugated system or the intensity of the acceptor and donor substituents), as well as its circumstances (such as the medium's polarity) (Chemla and Zyss 1987; Gorman and Marder 1993).

This variable is connected to a geometrical parameter in acceptor-donor polyenes, the bond length alternation (BLA) (Ahlheim et al. 1996), which is described as the median of the length alteration between adjacent carbon–carbon bonding in a polymethine $((CH)_n)$ ring. Polyenes have contrasting double and single and double bond lengths of 1.34 Å and 1.45 Å, respectively, resulting in a high level of BLA (+0.11 Å). The non-centrosymmetry needed for second-order NLO effect (such as the electro-optic impact) is introduced by directing dipolar NLO chromophores through applying an external EF to a poled polymer. Similarly, in a polymethine chain, the bond-order alternation (BOA) is measured on the basis of the variation in bond-order among adjacent carbon-carbon bonds. The BLA or BOA is found to be associated with hyper polarizability with polymethine as well as polyenes dyes (Prasad and Williams 1991).

5.12.3 Multiphoton Materials

Two-photon absorption (TPA) is a third-order mechanism in NLO that is connected with extended π-conjugated electronic structures and charge carrier properties, which can be measured ultimately by two-photon absorption radiation fluorescence (TPF). Organic composites that exhibit TPA show promising frequency up transition lasing, optical signal regulating, 3D fluorescence, 3D optical computing, 3D digital image microfabrication, and photothermal therapy. TPA is a phenomenon in which molecule absorbs two photons at the same time and is naturally weak at a usual intensity of light (Winter and Shroff 2014).

TPA permits for specific 3D spatial confinement of molecules during excitation. Since the rate of TPA is proportionate to the square of the strength of the occurring two-photon resonant radiation and the strength of a concentrated laser light reduces nonlinearly with distances from the central focus, TPA drops off as the fourth power of deviation from its focus. As a result, TPA along each focal plane is minimal (Sun and Kawata 2004). Furthermore, because the targeted molecules can be activated with higher wavelength well below those at which material demonstrates single-photon penetration, TPA can excite molecules at depths in a normally consuming medium (Winter et al. 2014).

Various ideas to manipulate TPA have also been presented as a result of these two considerations; moreover, for several applications, chromophores designed for the excitation of one photon were used, which had poor efficiency for the absorption of two photons, i.e., low TPA crossed section δ. TPA applications can become more efficient and usable because laser loss to molecules and nearby materials can be reduced and lower voltage, less costly lasers can be used (Tang et al. 2002).

5.13 CONCLUSION

Nonlinear effects and nonlinear optical SCs are of great importance especially in the field of photonics. Carbon-based compounds and chromophores are of great significance in field of NLO. Bulk organic SCs such as aromatic derivative Bz and m-DNB are reviewed briefly since these are used in NLO. Electro-optic polymers and multipole polymers are also discussed. ONL SCs have a variety of applications and will increase as more nonlinear aspects of SCs are explored experimentally and reviewed theoretically.

SHG has been used for biological visualization, and with some improvements, the SHG systems have been made to be appropriate for imaging collagen filaments in different organs. There is also another nonlinear multiphoton imaging method that is called THG, which appears to be more adaptable than SHG. In non-centrosymmetric framework, the second-order nonlinearity is the most reduced order of nonlinear interaction.

In NLO, optical properties of materials depend upon the intensity of incident light. NLO has been classified into two types: one is due to real and virtual absorption and the other is due to non-radiative recombination that's called photothermal optical nonlinearities. In photothermal nonlinearity, most of the part of absorbed light is transformed into heat. When the amount of heat is enough to enhance lattice temperature T_L, then optical properties of the material will change to give rise to nonlinearity. The PEONL are due to biexciton and first observed in the quantum well. There are three regions of the excitons: low density, intermediate density, and high density. The LEO effect arises from the non-centrosymmetric material's quadratic susceptibility $\chi^{(2)}(-\omega;\omega,0)$. When monochromatic light is exposed to a material then it gives rise to the material's optical frequency. The LEO effect is absent in the centrosymmetric materials. The RI dependency has been well explained via the utilization of the photorefractive effect. Nonlinear effects in SCs depend on the intensity of light. Nonlinear optical properties are dependent on the intensity of incident field. NLO is reviewed in the prospect of nonlinear susceptibility and how it gives rise to harmonic generation. Symmetry of the nonlinear susceptibility is discussed through relations derived from the fundamental Maxwell equations. General formulism is constructed for nonlinear optical properties of SCs by three regimes. Behavior of SCs and nonlinear effects in these three regimes are examined. Harmonic generation origin is reviewed and schemes of phase matching are reviewed.

REFERENCES

Agramovich, VM and Ginzburg, VL. 1984. Crystal Optics and Spatial Dispersion and Excitons, 2nd ed., Springer-Verlag, New York.

Ahlheim, Markus, Marguerite Barzoukas, Peter V Bedworth, Mireille Blanchard-Desce, Alain Fort, Zhong-Ying Hu, Seth R Marder, Joseph W Perry, Claude Runser, and Markus Staehelin. 1996. "Chromophores with strong heterocyclic acceptors: a poled polymer with a large electro-optic coefficient." *Science* 271 (5247):335–337.

K. Al-Hemyari, CN Ironside, and JS Aitchison. 1992. "Resonant nonlinear optical properties of GaAs-GaAlAs single quantum-well waveguide and an integrated asymmetric Mach-Zehnder interferometer." *IEEE Journal of Quantum Electronics* 28 (10):2051–2056.

Auston, DH, AM Glass, and AA Ballman. 1972. "Optical rectification by impurities in polar crystals." *Physical Review Letters* 28 (14):897.

Babu, G Anandha, R Perumal Ramasamy, Perumalsamy Ramasamy, and V Krishna Kumar. 2009. "Synthesis, crystal growth, and characterization of an organic nonlinear optical donor-π-acceptor single crystal: 2-amino-5-nitropyridinium-toluenesulfonate." *Crystal Growth and Design* 9 (7):3333–3337.

Barad, Y, Henryk Eisenberg, M Horowitz, and Yaron Silberberg. 1997. "Nonlinear scanning laser microscopy by third harmonic generation." *Applied Physics Letters* 70 (8):922–924.

Barvinschi, Floricica, Anca Stanculescu, and Florin Stanculescu. 2011. "Heat transfer process during the crystallization of benzil grown by the Bridgman–Stockbarger method." *Journal of Crystal Growth* 317 (1):23–27.

Boyd, Robert W. 2008. "The nonlinear optical susceptibility." *Nonlinear Optics* 3:1–67.

Broser, I, and J Gutowski. 1988. "Optical nonlinearity of CdS." *Applied Physics B* 46 (1):1–17.

Capper, Peter, Arthur Willoughby, and Safa Kasap. 2020. *Optical Properties of Materials and Their Applications*: John Wiley & Sons, New Jersey.

Chemla, Daniel S, David Miller, Peter Smith, Arthur Gossard, and William Wiegmann. 1984. "Room temperature excitonic nonlinear absorption and refraction in GaAs/AlGaAs multiple quantum well structures." *IEEE Journal of Quantum Electronics* 20 (3):265–275.

DS, Chemla, and Zyss J, Ed. 1987. *Nonlinear Optical Properties of Organic Molecules and Crystals*. Vol. 1: Orlando, FL, Academic Press.

Christodoulides, Demetrios N, Iam Choon Khoo, Gregory J Salamo, George I Stegeman, and Eric W Van Stryland. 2010. "Nonlinear refraction and absorption: mechanisms and magnitudes." *Advances in Optics and Photonics* 2 (1):60–200.

Condon, EU (1937) Theories of optical rotatory power, Rev. *Mod. Phys.* 9:432–457.

Drake, Gordon WF. 2006. *Springer Handbook of Atomic, Molecular, and Optical Physics*. Springer Science & Business Media, New York.

Freund, Isaac, Moshe Deutsch, and Aaron Sprecher. 1986. "Connective tissue polarity. Optical second-harmonic microscopy, crossed-beam summation, and small-angle scattering in rat-tail tendon." *Biophysical Journal* 50 (4):693–712.

Ganeev, Rashid 2005. "Nonlinear refraction and nonlinear absorption of various media." *Journal of Optics A: Pure and Applied Optics* 7 (12):717.

Giordmaine, Joseph Anthony, and Robert C Miller. 1965. "Tunable coherent parametric oscillation in LiNbO$_3$ at optical frequencies." *Physical Review Letters* 14 (24):973.

Gopalakrishnan, Srinivasan, and Saggam Narendar. 2013. *Wave Propagation in Nanostructures: Nonlocal Continuum Mechanics Formulations*. Springer Science & Business Media, Switzerland.

Gorman, Christopher B, and Seth R Marder. 1993. "An investigation of the interrelationships between linear and nonlinear polarizabilities and bond-length alternation in conjugated organic molecules." *Proceedings of the National Academy of Sciences* 90 (23):11297–11301.

Gunshor, Robert L, and Leslie A Kolodziejski. 1988. "Recent advances in the molecular beam epitaxy of the wide-bandgap semiconductor ZnSe and its superlattices." *IEEE Journal of Quantum Electronics* 24 (8):1744–1757.

Haug, Hartmut, and Stefan Schmitt-Rink. 1984. "Electron theory of the optical properties of laser-excited semiconductors." *Progress in Quantum Electronics* 9 (1):3–100.

He, Guang Sheng, and Song-hao Liu. 1999. *Physics of Nonlinear Optics*. World Scientific Publishing Company.

K-H, Hellwege Ed. 1982. *Landolt–Börnstein, New Series*. New York, Springer Verlag.

Henneberger, K, and H Haug. 1988. "Nonlinear optics and transport in laser-excited semiconductors." *Physical Review B* 38 (14):9759.

Hobden, MV. 1967. "Phase-matched second-harmonic generation in biaxial crystals." *Journal of Applied Physics* 38 (11):4365–4372.

Hönerlage, Bernd, Roland Levy, Jean Bernard Grun, Claus Klingshirn, and Klaus Bohnert. 1985. "The dispersion of excitons, polaritons and biexcitons in direct-gap semiconductors." *Physics Reports* 124 (3):161–253.

Hulin, D, A Mysyrowicz, A Antonetti, Arnold Migus, WT Masselink, H Morkoc, HM Gibbs, and Nasser Peyghambarian. 1986. "Well-size dependence of exciton blue shift in GaAs multiple-quantum-well structures." *Physical Review B* 33 (6):4389.

Ironside, Charlie N. 1993. "Optical nonlinear effects in semiconductors." In *Principles and Applications of Nonlinear Optical Materials*, 35–75: Springer, Dordrecht.

Kleinman, David A. 1962a. "Nonlinear dielectric polarization in optical media." *Physical Review* 126 (6):1977.

Kleinman, David A. 1962b. "Theory of second harmonic generation of light." *Physical Review* 128 (4):1761.

Klingshirn, Claus. 1990. "Non-linear optical properties of semiconductors." *Semiconductor Science and Technology* 5 (6):457.

Klingshirn, Claus, and Hartmut Haug. 1981. "Optical properties of highly excited direct gap semiconductors." *Physics Reports* 70 (5):315–398.

Klingshirn, C, Ch Weber, HE Swoboda, R Renner, FA Majumder, M Kunz, M Rinker, Holger Schwab, M Wegener, and DS Chemla. 1989. "Photo-electronic optical nonlinearities in three-and quasi two-dimensional semiconductors." Proc. SPIE 1017, *Nonlinear Optical Materials*, (8 March 1989); https://doi.org/10.1117/12.949952.

Koch, Stephan W, Nasser Peyghambarian, Markus Lindberg, and BD Fluegel. 1989. "Femtosecond dynamics of semiconductor nonlinearities: theory and experiments." In *Optical Switching in Low-Dimensional Systems*, 139–150: Springer, Boston, MA.

Krowne, Clifford M. 2019. "Introduction to examination of 2D hexagonal band structure from a nanoscale perspective for use in electronic transport devices." In *Advances in Imaging and Electron Physics*, 1–6: Elsevier.

Kumar, Vikas, Michele Casella, Egle Molotokaite, Davide Gatti, Philipp Kukura, Cristian Manzoni, Dario Polli, Marco Marangoni, and Giulio Cerullo. 2012. "Balanced-detection Raman-induced Kerr-effect spectroscopy." *Physical Review A* 86 (5):053810.

Kumar, Vikas, Nicola Coluccelli, Marco Cassinerio, Michele Celebrano, Abigail Nunn, Massimo Levrero, Tullio Scopigno, Giulio Cerullo, and Marco Marangoni. 2015. "Low-noise, vibrational phase-sensitive chemical imaging by balanced detection RIKE." *Journal of Raman Spectroscopy* 46 (1):109–116.

Kumar, Vikas, Nicola Coluccelli, and Dario Polli. 2018. "Coherent optical spectroscopy/microscopy and applications." *Elsevier Science BV*:87–115.

Landau, LD, and EM Lifshitz. 1984. *Electrodynamics of Continuous Media* (revised and enlarged, with Pitaevskii, LP). Oxford, Pergamon Press.

Landau, Lev Davidovich, JS Bell, MJ Kearsley, LP Pitaevskii, EM Lifshitz, and JB Sykes. 2013. *Electrodynamics of Continuous Media*. Vol. 8: Elsevier.

Marder, Seth R. 2006. "Organic nonlinear optical materials: where we have been and where we are going." *Chemical Communications* Vol.2, 131–134.

Marder, Seth R., David N. Beratan, and L-T Cheng. 1991. "Approaches for optimizing the first electronic hyperpolarizability of conjugated organic molecules." *Science* 252 (5002):103–106.

Matteini, Paolo, Fulvio Ratto, Francesca Rossi, Riccardo Cicchi, Chiara Stringari, Dimitrios Kapsokalyvas, Francesco S Pavone, and Roberto Pini. 2009. "Photothermally-induced disordered patterns of corneal collagen revealed by SHG imaging." *Optics Express* 17 (6):4868–4878.

McGill, TC, Clivia M. Sotomayor Torres, and W. Gebhardt. 2012. *Growth and Optical Properties of Wide-Gap II–VI Low-Dimensional Semiconductors*. Vol. 200: Springer Science & Business Media.

Meyers, Freel, Seth R. Marder, Brian M. Pierce, and Jean-Luc Bredas. 1994. "Electric field modulated nonlinear optical properties of donor-acceptor polyenes: sum-over-states investigation of the relationship between molecular polarizabilities (. alpha.,. beta., and. gamma.) and bond length alternation." *Journal of the American Chemical Society* 116 (23):10703–10714.

Munn, Robert W, and Charlie N Ironside. 1993a. *Principles and Applications of Nonlinear Optical Materials*. Springer, Dordrecht.

Robert W. Munn, and Ironside Charlie, Eds. 1993b. *Principles and Applications of Non-linear Optical Materia*. Springer, Dordrecht.

Oudar, Jean-Louis, and Daniel S Chemla. 1977. "Hyperpolarizabilities of the nitroanilines and their relations to the excited state dipole moment." *The Journal of Chemical Physics* 66 (6):2664–2668.

Peceli, Davorin. 2013. "Absorptive and refractive optical nonlinearities in organic molecules and semiconductors." *Electronic Theses and Dissertations*. 2571.

Plotnikov, Sergey V, Andrew C Millard, Paul J Campagnola, and William A Mohler. 2006. "Characterization of the myosin-based source for second-harmonic generation from muscle sarcomeres." *Biophysical Journal* 90 (2):693–703.

Prasad, Paras N, and David J Williams. 1991. *Introduction to Nonlinear Optical Effects in Molecules and Polymers*. Vol. 1: New York, Wiley.

Rao, A Srinivasa. 2016. "Overview on Second and Third Order Optical Nonlinear Processes." *arXiv preprint arXiv:1612.09399*.

Rice, A, Y Jin, XF Ma, X-C Zhang, David Bliss, J Larkin, and M Alexander. 1994. "Terahertz optical rectification from <110> zinc-blende crystals." *Applied Physics Letters* 64 (11):1324–1326.

Rohleder, Józef Władysław, and Robert W Munn. 1992. *Magnetism and Optics of Molecular Crystals*. John Wiley & Sons Incorporated, New Jersey.

Schmitt-Rink, S, DS Chemla, and David AB Miller. 1985. "Theory of transient excitonic optical nonlinearities in semiconductor quantum-well structures." *Physical Review B* 32 (10):6601.

Schmitt-Rink, S, DS Chemla, and David AB Miller. 1989. "Linear and nonlinear optical properties of semiconductor quantum wells." *Advances in Physics* 38 (2):89–188.

Schöll, Eckehard, and Edith Scholl. 2001. *Nonlinear Spatio-Temporal Dynamics and Chaos in Semiconductors*: Cambridge University Press.

Squier, Jeff A, Michiel Müller, GJ Brakenhoff, and Kent R Wilson. 1998. "Third harmonic generation microscopy." *Optics Express* 3 (9):315–324.

Stanculescu, Anca. 2007. "Investigation of the growth process of organic/inorganic doped aromatic derivates crystals." *Journal of Optoelectronics and Advanced Materials* 9 (5):1329.

Stanculescu, Anca, Laura Tugulea, Horia V Alexandru, Florin Stanculescu, and Marcela Socol. 2005. "Molecular organic crystalline matrix for hybrid organic–inorganic (nano) composite Materials." *Journal of Crystal Growth* 275 (1–2):e1779–e1786.

Stanculescu, Anca, and Florin Stanculescu. 2015. "Organic semiconductors for non-linear optical applications." In *Optoelectronics-Materials and Devices*: IntechOpen, Croatia.

Stanculescu, Anca, Loredana Vacareanu, Mircea Grigoras, Marcela Socol, Gabriel Socol, Florin Stanculescu, Nicoleta Preda, Elena Matei, Iulian Ionita, and Mihaela Girtan. 2011. "Thin films of arylenevinylene oligomers prepared by MAPLE for applications in non-linear optics." *Applied Surface Science* 257 (12):5298–5302.

Stanculescu, Florin, Iulian Ionita, and Anca Stanculescu. 2014. "Organic/inorganic-doped aromatic derivative crystals: growth and properties." *Journal of Crystal Growth* 401:215–220.

Sun, Chi-Kuang, Shih-Wei Chu, Shi-Peng Tai, Stacia Keller, Umesh K Mishra, and Steven P DenBaars. 2000. "Scanning second-harmonic/third-harmonic generation microscopy of gallium nitride." *Applied Physics Letters* 77 (15):2331–2333.

Sun, Hong-Bo, and Satoshi Kawata. 2004. "Two-photon photopolymerization and 3D lithographic microfabrication." In *NMR 3D Analysis Photopolymerization*, 169–273: Springer.

Tang, Xin-Jing, Li-Zhu Wu, Li-Ping Zhang, and Chen-Ho Tung. 2002. "Two-photon-pumped frequency-upconverted lasing and optical power limiting properties of vinylbenzothiazole-containing compounds in solution." *Physical Chemistry Chemical Physics* 4 (23):5744–5747.

Tripathi, Ashutosh Kumar, Michael Heinrich, Theo Siegrist, and Jens Pflaum. 2007. "Growth and electronic transport in 9, 10-diphenylanthracene single crystals—an organic semiconductor of high electron and hole mobility." *Advanced Materials* 19 (16):2097–2101.

Tsang, Thomas YF. 1995. "Optical third-harmonic generation at interfaces." *Physical Review A* 52 (5):4116.

Winter, Peter W, and Hari Shroff. 2014. "Faster fluorescence microscopy: advances in high speed biological imaging." *Current Opinion in Chemical Biology* 20:46–53.

Winter, Peter W, Andrew G York, Damian Dalle Nogare, Maria Ingaramo, Ryan Christensen, Ajay Chitnis, George H Patterson, and Hari Shroff. 2014. "Two-photon instant structured illumination microscopy improves the depth penetration of super-resolution imaging in thick scattering samples." *Optica* 1 (3):181–191.

Yan, RH, Scott W Corzine, Larry A Coldren, and I Suemune. 1990. "Corrections to the expression for gain in GaAs." *IEEE Journal of Quantum Electronics* 26 (2):213–216.

Yelin, Dvir, Yaron Silberberg, Yaniv Barad, and Jay S Patel. 1999. "Depth-resolved imaging of nematic liquid crystals by third-harmonic microscopy." *Applied Physics Letters* 74 (21):3107–3109.

Yin, M, HP Li, SH Tang, and Wenhai Ji. 2000. "Determination of nonlinear absorption and refraction by single Z-scan method." *Applied Physics B* 70 (4):587–591.

Zimmermann, Roland. 1988. *Many-Particle Theory of Highly Excited Semiconductors*. Vol. 18: Wiley-VCH Pub.

6 Semiconductor Photoresistors

Anoop Singh, Aamir Ahmed, and Sandeep Arya

CONTENTS

6.1 INTRODUCTION

The word "photoresistor" is made of two words "photons" and "resistor". The photoresistor is a special kind of resistor that is used in electronic circuits. The photoresistor may be defined as a resistor, the resistance of which depends upon the intensity of light incident upon it. The resistance of the photoresistor changes with the intensity of light. It comprises a conductor material that is sensitive towards light, i.e., photo-conductive material. The phenomenon governing photoresistor is very simple, light is made incident on a photoresistor, and excitation of electrons takes place from the valence to conduction band. More is the number of electrons transferred towards the conduction band, less will be the resistance offered by the photoresistor. The material that is most commonly used for fabricating photoresistors is a highly resistive semiconductor. In terms of conductivity, the materials are characterized as conductors, semiconductors, and insulators. In a conductor, the electrons can move on applying a potential across the conductor, whereas in an insulator, the electrons cannot move freely [1]. The semiconductors have attained huge interest in recent times as it has the conductivity in between the insulators and conductors. It bridges the gap between a conductor and an insulator. In a semiconductor, there exist energy bands that are formed by the close arrangement of similar energy levels of electrons. The energy band having the lowest energy is called the valence band and the energy band having the highest energy is called the conduction band. The energy difference between the valence and conduction band is called the energy bandgap. More are the number of electrons in the conduction band of a semiconductor, more conducting will be the semiconductor. Thus, the resistance offered by the semiconductor towards the flow of electrons depends upon the number of electrons in the conduction band. In the case of a photoresistor, the light incident on it excites the electrons from the valence

DOI: 10.1201/9781003188582-6

band to the conduction band by providing them the energy to cross the energy bandgap barrier. Since upon illumination, the number of electrons in the conduction band increases, the resistance of the photoresistor decreases. This change in the resistance of the photoresistor can be measured and used for driving various electronic components. For example, during the night when the light decreases, the photoresistor can be used to turn on the light. It can also be used as a simple light detector. Due to its potential applications in automation, optoelectronic devices, monitoring processes, and chemical and biological analysis [2], photoresistors have gained tremendous interest in the last few decades. The photoresistors are categorized into three categories based on their spectral properties, i.e., IR-sensitive photoresistors, UV-sensitive photoresistors, and visible-light-sensitive photoresistors. The materials like CdS and CdSe have been successfully used as visible-light-sensitive photoresistors [3, 4], whereas PbS and PbSe have been used as IR-sensitive photoresistors [5, 6]. The zinc oxide (ZnO) due to its wide bandgap of about 3.37 eV has been developed as a UV-sensitive photoresistor [7]. In most photo-resistors, the photoconduction process is responsible for the resistance change upon illumination. The photoconduction is a fast process due to which such photoresistors exhibit fast detection, i.e., quick recombination or trapping of the generated electron-hole pairs. In the case of some photoresistors such as n-type ZnO, the change in the resistance is due to effects taking place at the surface of the material. In such materials, the sensitivity towards illumination is quite low [8]. The surface effects that mostly take place in such materials are the adsorption and desorption of the hydroxyl and oxygen on the surface of the material, which results in resistance change. But there is always a scope for the development and enhancement of such materials. By increasing the surface-to-volume ratio in such kinds of materials, the sensitivity towards light can be increased. The nanorods of ZnO have shown better sensitivity as photoresistors as compared to their bulk counterparts [9]. The switching in the resistance of the photoresistors can be used for the development of next-generation and highly effi-cient resistance random-access memory (RRAM). These RRAMs are strong, fast to write and erase, and save power and energy [10]. The resistance switching has been successfully depicted by some of the materials such as graphene oxide, amorphous carbon, metal oxides of transition elements, and perovskites [11–14]. The resistance change can also be depicted by drifting of the ions upon the fabri-cation or breakage of the conducting filaments, trapping of the electron-hole pairs, and a special kind of tunnelling called Fowler-Nowler tunnelling [15–17]. Another nanomaterial that has gained huge interest in recent times is the carbon nanotube (CNT). The characteristics of CNTs such as photolumi-nescence and conductivity have been studied vastly [18]. They have been found to possess an efficient mechanism of electron transport and their potential for electrical and photonic applications has been explored [19–22]. The CNTs possess a different density of states as compared to other materials, thus producing electron-hole pairs upon irradiation with different wavelengths. Moreover, the bandgap of the CNTs can be tuned by simply adjusting their carrier mobility and diameter [23]. Thus, they possess huge potential to be used for photoresistor applications such as wearable photoresistor devices [24].

6.2 TYPES OF PHOTORESISTORS

Based on the material used for fabricating a photoresistor, these have been classified into two types, i.e., intrinsic photoresistor and extrinsic photoresistor.

6.2.1 INTRINSIC PHOTORESISTOR

As the word "intrinsic" suggests pure or free from impurities, the intrinsic photoresistors possess the semiconductor material that has not been doped or simply the pure semiconductor material. In such photoresistors, electrons in the valence band are excited to the conduction band of semiconduc-tor upon illumination. Among various intrinsic semiconductors that have been used for fabricating photoresistors, germanium (Ge) has been widely used. The Ge possesses an indirect bandgap of 0.66 eV and has been previously used for preparing photo and radiation detectors, tracking photons and charged particles, and IR-sensitive photoresistors [25–30]. In the last few decades, the growing interest in the field of nanoscience and nanotechnology has led to the development of quantum dots and nanowires of the Ge and Si [31]. As we already know that, at the nanoscale, the properties of the material are enhanced as compared to its bulk counterparts; the nanomaterials of Ge and Si have been used to develop more efficient devices, sensors, and detectors. The proper alignment of the nanowires of Ge and Si is necessary for successfully using these nanomaterials for electronic applications. To obtain this, anodized aluminium oxide (AAO) templates or mesoporous materials are used to deposit the nanowires [32, 33]. Recently, there has been tremendous development in the design and structure of the AAO templates. The AAO templates have been deposited on silicon substrates [34]. Such templates have been found to have ordered pores that are very beneficial in semiconductor electronics. It is important to mention here that the luminescence and electrical properties of the simple semiconductor nanowires have been investigated, but no significant breakthrough has been made in studying the electrical and optical properties of the aligned nanowires [18]. So, there is a scope for the investigation of the photoluminescence and electrical conductivity of the ordered semiconductor nanowires along with their applications as photoresistors. An attempt has already been made in this regard, where the ordered nanowires of Ge have been deposited on the AAO template [35].

6.2.2 Extrinsic Photoresistor

The word "extrinsic" suggests here some external species or entities. The extrinsic photoresistors contain some impurity or dopant that has been incorporated into the semiconductor material. The doping of the impurity results in enhancing the photo-conductive behaviour of the semiconductor material. The impurity selected for doping is decided such that it does not get ionized during illumination with light. During conduction in the extrinsic photoresistors, there occurs the excitation of the charges between the impurity and conduction or valence band of the semiconductor material. Various extrinsic photoresistors have already been fabricated. The ZnO as already discussed has a wide bandgap and is commonly used for photoresistor applications [36]. The limitation of using ZnO as a photoresistor is that it is only responsive towards UV light and insensitive towards visible light. To make it responsive towards visible light, it has been prepared as a junction with different semiconductor materials such as ZnO-Si junction [37]. The ZnO-Si junction material acts as a diode and shows resistance change upon illumination with visible light due to current flow through the diode. Zhang et al. [38] fabricated ZnO nanorods and Si junction material, i.e., ZnO (NRs)-Si, which shows resistance change upon illumination with visible light. Both the schematic layout scheme and energy band diagram of the photoresistor are shown in Figure 6.1(a and b). However, in this case, the current flow is not through the diode but due to Si present in the material. However, the presence of

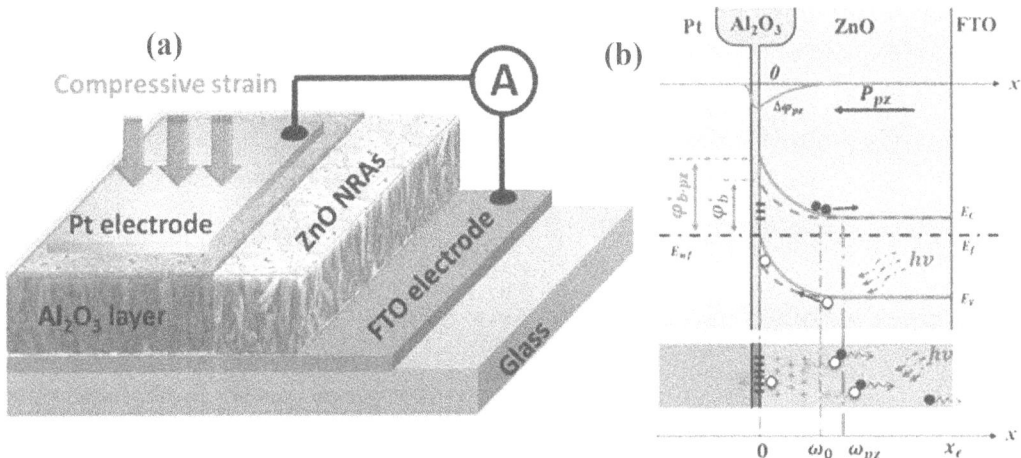

FIGURE 6.1 (a) Schematic scheme of the photoresistor and (b) energy band diagram of the photoresistor. (Reproduced with permission from [38]. Copyright 2014, Elsevier.)

ZnO NRs helps in collecting light, increasing the thickness of the depletion layer in Si, and creating a channel for the transfer of photogenerated charges. The photoresistor was found to be sensitive, fast, and independent of the gaseous environment. The photoresistor based on single walled carbon nanotubes (SWCNT) systems is also attracting much attention of the researcher [39, 40]. Figure 6.2(a) diagrammatically represents SWCNT made of pristine, Figure 6.2(b) shows boron-doped SWCNT and Figure 6.2(c) shows the scattering region of nitrogen-doped SWCNT [41].

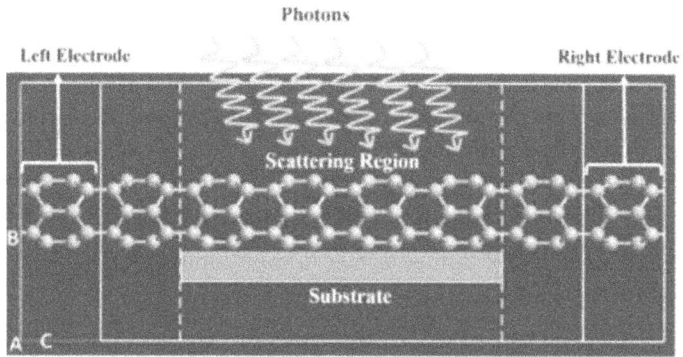

(a)

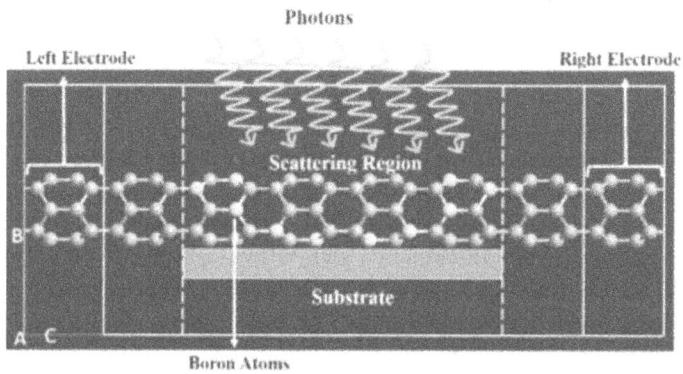

(b)

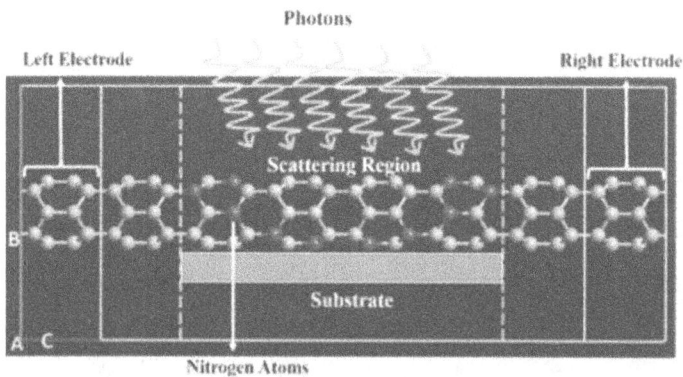

(c)

FIGURE 6.2 Schematic diagram of (a) pristine, (b) boron-doped, and (c) nitrogen-doped SWCNT systems. (Reproduced with permission from [41]. Copyright 2020, Elsevier.)

6.3 WORKING PRINCIPLE OF SEMICONDUCTOR PHOTORESISTOR

The discovery of photoconductivity in the selenium (Se) led to the discovery of various new materials having similar properties. In the 1930s and 1940s, followed by the development of Si and Ge photoconductors, lead sulphide (PbS), lead selenide (PbSe), and lead telluride (PbTe) were the other materials used for photoconductor applications. In modern technology, light-responsive resistors (photoresistors) are mostly made up of lead sulphide, lead selenide, indium antimonide, cadmium sulphide, and cadmium selenide. Among them, the photoresistor made of CdS has gained much attention due to its effective photo-resistive properties. To fabricate a CdS photoresistor, CdS powder is mixed with a binding material (inert) and the mixture is then subjected to high pressure followed by the sinteration. Using the phenomenon of vacuum evaporation, the mixture is then deposited on one side of a substrate in the form of electrodes. The disc formed is prevented from surface contamination by encapsulation in plastic or mounting it in an envelope made of glass. The CdS is very suitable for fabricating photoresistors because its spectral response curve is similar to a human eye. Moreover, the wavelength of peak sensitivity for CdS is in the range 500–600 nm, lying in the visible region. Many countries in the world do not support the application of Cd or Pb due to their toxicity and hence, in such countries, the photoresistors made of CdS or PbS or PbSe are not used or fabricated. Since the valence electrons are loosely bound to the nucleus and are present in the outermost shell of an atom. A small amount of energy is enough to knock out these electrons from the atom. When the required energy is supplied to these valence electrons, they become free and are no longer attached to the nucleus. These electrons move freely under the influence of an external electric field. The energy of light can be used to remove valence electrons from an atom in some materials. This acts as the working principle of a photoresistor. The light incident on the material is absorbed, creating free electrons. The semiconductors with no PN junction are mostly preferred for fabricating photoresistors [42].

6.4 CHARACTERISTICS OF SEMICONDUCTOR PHOTORESISTOR

6.4.1 WAVELENGTH DEPENDENCY

The wavelength of the light used to illuminate a photoresistor is very important. The sensitivity of a photoresistor depends upon the wavelength of light and a photoresistor works only in a certain wavelength range. Beyond this range, there will be no change in the resistance of the photoresistor. This wavelength range varies from one material to another. When a response curve is plotted for wavelength against sensitivity, each material shows different behaviours from the other. Tsybrii-Ivasiv et al. [43] successfully measured the dependency of photoresistors on the wavelength of the incident light. The wavelength range is maximum for the extrinsic photoresistors, ranging towards the infrared region of the spectrum. But proper care must be taken while working with the photoresistors in the IR region because the measurements could be affected by the dependence of resistance on heat.

6.4.2 DEPENDENT ON SENSITIVITY

In comparison to phototransistors and photodiodes, the photoresistors have low sensitivity. The phototransistors and photodiodes possess a PN junction and act as active components, whereas the photoresistors lack PN junction and are passive components. The resistance in the photoresistors also depends upon the temperature that is observed when there occurs a change in the resistance at a constant intensity of light. As a result, the photoresistors cannot be used for measuring the intensity of light. Lin et al. [44] dipped a polyimide photoresistor in a solution of magnesium chloride $(MgCl_2)$ and studied the effect of inorganic salt on the sensitivity and efficiency of the photoresistor. The concentration of the salt and the time of dispersion are the two main factors affecting the sensitivity and efficiency of the photoresistor. The EDS spectrum confirmed the presence of $MgCl_2$ on the photoresistor layer. Moreover, the photoresistor was found to have high sensitivity and can be used as a humidity sensor. From Figure 6.3, it is observed that with the variation in the doping criteria, the sensitivity of the developed sensors is also changed [44].

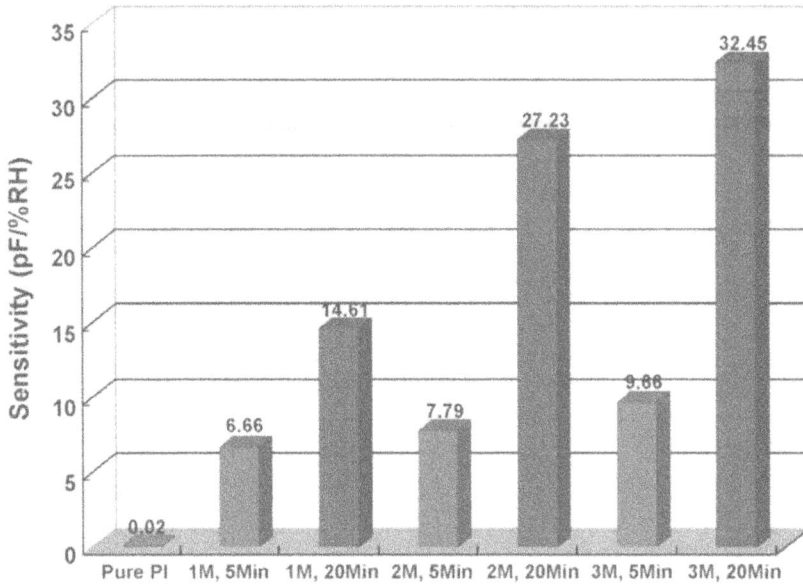

FIGURE 6.3   Relative sensitivity of the sensors prepared using different doping criteria. (Reproduced with permission from [44]. Copyright 2008, Elsevier.)

6.4.3   LATENCY DEPENDENT

The photoresistors possess a unique property called resistance recovery rate, i.e., there is a time lag between the irradiation and resistance change. The irradiation and resistance change are not simultaneous in the photoresistors; there is a time lag between the two phenomena. On the one hand, when the light is incident on a photoresistor, the resistance takes about 10 μs to drop completely. On the other hand, the resistance takes about 1 s to get to the original value when the light is completely removed. Due to this reason, the photoresistors are not suitable for applications requiring quick resistance change, i.e., recording light fluctuations and light control. But this time lag in the illumination and resistance change can be used in other applications such as audio compressors.

6.5   APPLICATIONS OF SEMICONDUCTOR PHOTORESISTOR

6.5.1   LIGHT SENSOR

Most of the light sensors use photoresistors. The photoresistors are quite efficient in recording the intensity of light and detecting the presence of light. For example, night lights and light meters are used in photography. The emerging field of photoresistor is optical biosensing. It can be performed while utilizing the photoresistor as an optical biosensor. An optical biosensor is a kind of biosensor in which the light emitted or absorbed by the sample is used for the sensing process. The optical biosensors and electrochemical biosensors have been developing in parallel to each other. But earlier, the optical biosensors were only available in the form of test strips showing colour change [45]. The different optical properties that are used in optical biosensing are absorption, emission, fluorescence, refractive index, phosphorescence, etc. [46]. The main benefits of using optical biosensors are that they do not require any reference sensor and no electric current is required for sensing. The optical biosensor based on the absorption of light is most commonly used due to its simple design and easy operation. Absorption is a phenomenon where light is absorbed by a material resulting in the excitation of electrons in an atom to the excited state. Since the lifetime of electrons in the excited state is very small (μs), it eventually drops to the ground state and releasing the excess energy non-radiatively.

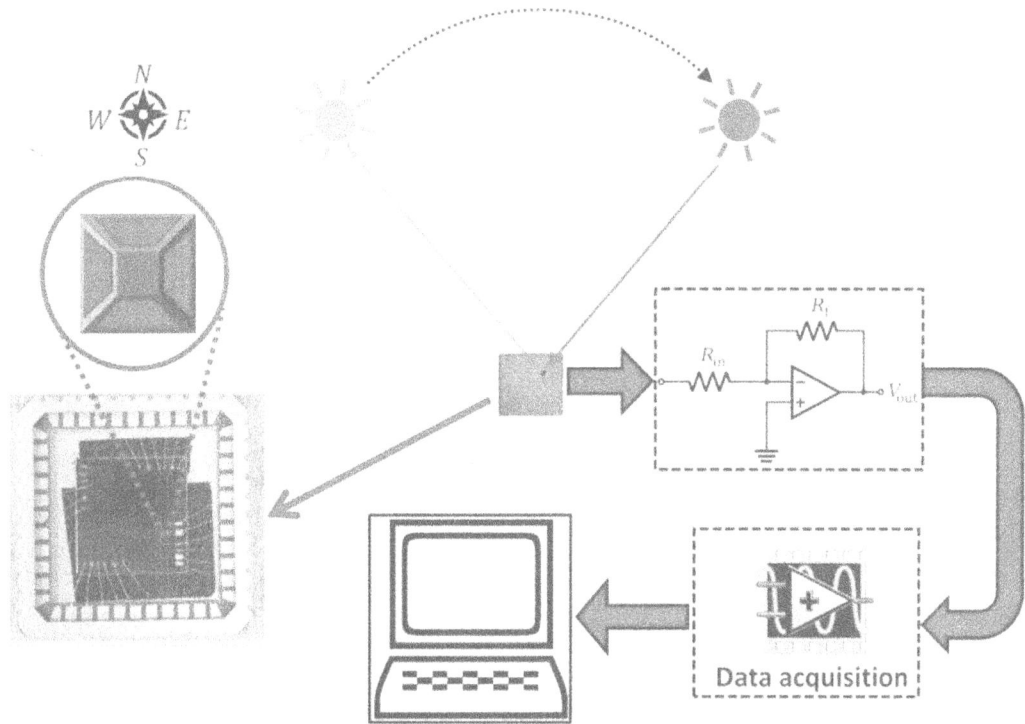

FIGURE 6.4 The test set-up for packaged devices. The cathode of each photodiode is connected to a trans-impedance amplifier where their common anode is connected to a slight negative bias if −1 V. Signals from the photodiodes are then captured and processed using the described algorithm. (Reproduced with permission from [49]. Copyright 2020, Elsevier.)

The Beer-Lambert Law is very important in the absorption phenomenon as it relates the absorbance to the concentration and needs to be considered before the absorption process.

The design of an optical biosensor consists of a source of light (an LED), a medium to transmit the optical signal (waveguide, optical fibre, etc.), a light detecting system, and a biological entity that can detect the changes (microorganism, enzyme, antibody). The photoresistors are used for detecting the optical signal as they are highly sensitive to light. The optical biosensors are most commonly used for detecting the glucose level in blood [47, 48]. Zhuo et al. [49] developed a micromachined vector light sensor in which the negative terminal of each photodiode was connected to an amplifier (trans-impedance). The whole set-up is demonstrated in Figure 6.4. Dutta et al. [50] fabricated a paper-based sensor to monitor the level of amylase in the human serum. The step-by-step fabrication of the sensor is shown in Figure 6.5. First, the surface of a paper was coated with starch (Prussian blue) and a solution of iodine. On the coated paper, the solution of amylase was dispensed, which resulted in the fading of blue colour due to the hydrolysis of the starch molecules. Following this, an LED was used to illuminate the surface of the paper. The light reflected from the paper was detected using a photoresistor. The resistance of the photoresistor was found to be sensitive towards the colour change and concentration of the amylase. The paper-based sensor showed remarkable sensitivity towards the detection of amylase at different concentrations and can be used for detecting amylase in the human blood serum.

6.5.2 Audio Compressors

An audio compressor is a device that is used to reduce the gain of an audio amplifier when it goes beyond a particular set value. The audio compressors are used to prevent the clipping of loud sounds

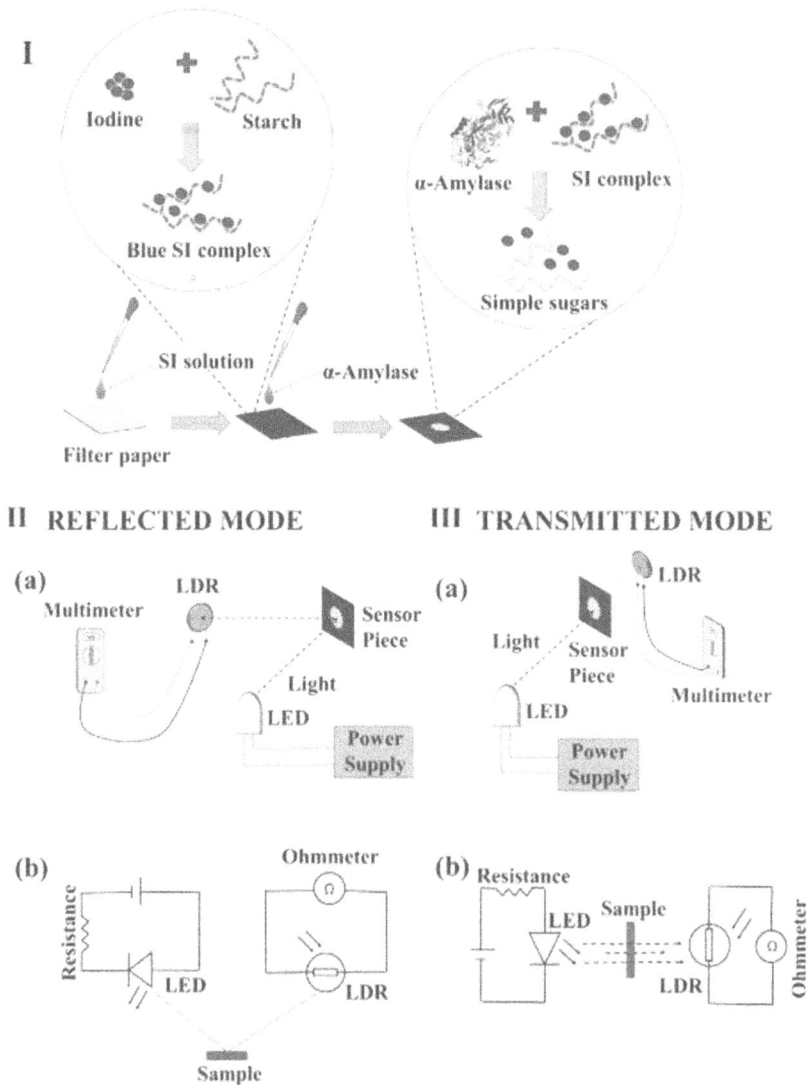

FIGURE 6.5 (I) Schematic illustrations of the steps to fabricate paper sensor. (II) Reflection mode – images (a) and (b) show the experimental set-up and corresponding circuit diagram for resistance measurement. (III) Transmitted mode – images (a) and (b) show the experimental set-up and the corresponding circuit diagram for resistance measurement. The notation LDR (LED) represents photoresistor or photodetector. (Reproduced with permission from [50]. Copyright 2016, Elsevier.)

and amplify the soft sounds at the same time. The photoresistors are used in an audio compressor along with an LED to control the gain of the compressor. As we already know, there is a time lag between the illumination and resistance change in a photoresistor, which helps in smoothing the signal characteristics. Moreover, the delay time of the photoresistor to be used in an audio compressor is about 0.1 s [51].

6.5.3 MEASUREMENT OF INCIDENT LIGHT INTENSITY

Thom et al. [52] successfully measured the intensity of the light produced by an LED using a photoresistor. It was found that the intensity of light emitted is inversely proportional to the resistance of the photoresistor. Figure 6.6 shows the measurement of the intensity of light from an LED source.

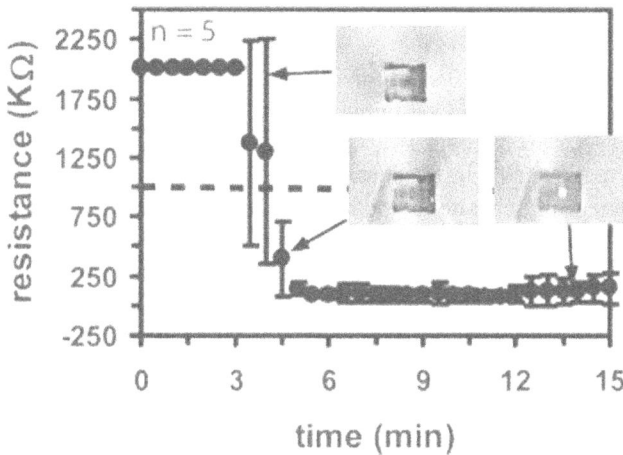

FIGURE 6.6 A photoresistor was used to measure the intensity of light emitted by the red LED after a 10-mL sample of water was added to the device. The intensity of light is approximately inversely proportional to the resistance measured with the photoresistor. The LED is illuminated at all points below the dotted line in (d). The measurements in (d) were repeated five times. (Reproduced with permission from [52]. Copyright 2012, Royal Society of Chemistry.)

6.5.4 LIGHT CONTROL CIRCUIT

In a photoresistor, the resistance varies with the intensity of light and becomes less with an increase in the intensity and vice versa. The photoresistor can be used to convert light into electric signals when a voltage is applied across the photoresistor. At a particular wavelength, the current increases with an increase in the intensity of light. Thus, the current produced can be controlled by controlling the optical signal. Wang designed a switch circuit using a photoresistor that can be controlled via an optical signal [53]. In a dark environment, the resistance is high, whereas in a bright environment, the resistance is low. Moreover, the light control unit can work both with light and sound [42]. The de-tuning of parallel resonant circuits like the receiver coil of magnetic resonance imaging (MRI) can be done using the photoresistor [54]. The photoresistor is connected in parallel with the capacitor and inductor of the receiver coil. An optical fibre is used to connect the photoresistor to a light source. When the light is made incident on the photoresistor, there occurs a resistance change that eventually decreases the Q value of the parallel circuit, thus achieving the de-tuning of the parallel resonant circuit. The optically detuned circuit thus created can also be used to develop an active tracker by simply integrating it with the standard coil of an MRI. This helps in reducing the number of electrical connections, which protects the patient from risk and shocks. Saravanan et al. [55] designed an effective solar panel that can sense the position of the sun and effectively converts more solar energy into electrical energy. They used photoresistors as sensors, which detect the position of the sun, and the whole solar panel is moved using a motor in a particular direction. The idea is simple; instead of taking earth as the source of reference, they took the sun as the reference source. This design can generate about 30% more energy as compared to traditional solar panels.

6.5.5 AUTOMATIC STREET LIGHTING

Street lighting in the majority of the world is still based on traditional methods and techniques. The goals of sustainable development and climate change have led the researchers to think of new materials and techniques that can be used to cut the emission of CO_2 in the air and deposition of hazardous material into the environment. To overcome these issues, many solutions have been formulated from time to time. Using LEDs that consume less power is one of the solutions, whereas controlling the lightning systems remotely is another solution formulated. The third solution is the use of solar energy, a renewable source of energy, for powering these street lights. Subramanyam et al. [56] claimed that developing a system comprising all three aspects has been formulated. It uses solar energy to light the street lamps containing LEDs as a light source via a control unit system. The system works on solar energy, thus reducing extra wires, costs, and saving energy. The control unit gathers the information and transfers it using a wireless system called the ZigBee protocol. The lighting system automatically operates at the night with the help of sensors and uses cost-effective LEDs. The system saves a lot of energy because it does not operate unnecessarily and all the components are turned off in the daylight. The battery used to store the charge depends upon the environmental condition under which an automatic street light is to be used. The photoresistors are used in the solar panels to effectively collect solar energy. Wang et al. [57, 58] designed street lights so that they can switch on just before the sun sets and switch off with sufficient light on the streets. The motion sensors are used to control the intensity of the light.

6.5.6 Wearable Image Sensing Applications

In the last few years, there has been a lot of work on the designing of wearable electronic devices. Wearable image sensing is one of the applications of these wearable devices that can be used for the biomedical purpose [59, 60]. An image sensor usually consists of a photosensitive device and a unit that can process these signals. The photoresistors are highly sensitive towards light and show a change in their resistance with variation in the intensity of the light. Hence, these can be used in image sensing devices. Yang et al. [61] fabricated a wearable sensor including a photoresistor device as shown in Figure 6.7. Complementary metal-oxide semiconductors (CMOS) are mostly used for signal processing in an image sensor [62]. But the problem with these processing units is that they are not flexible and cannot be used for wearable image sensing. However, RRAM made from parylene-C is flexible and can be used for image sensing. The photo-resistive RRAM may be used as a portable image sensor [63]. The main benefit of using this device for image detection is that it can be read and deleted very easily and several times without affecting performance.

6.5.7 Photogate Timing with a Smartphone

The information about the temperature can be easily obtained by a circuit consisting of a thermistor that is connected to mobile via headset jack [64]. In the same circuit, the thermistor can be replaced and the set-up can be used to perform various functions and measurements. When the thermistor is replaced by a photoresistor, the signal amplitude will change according to the resistance change in

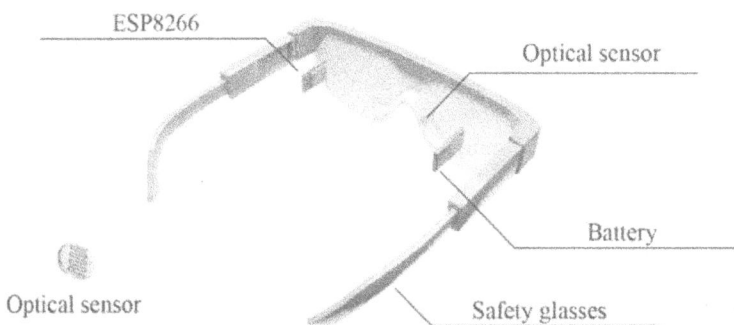

FIGURE 6.7 The sketch of safety glasses for wearing detection with an ESP8266 chip embedded in left temple and a photoresistor in a nose pad. (Reproduced with permission from [61]. Copyright 2020, Elsevier.)

the photoresistor. The signal does not drop instantly when the resistor is blocked but rather tapers off to a minimum while being blocked. Also, due to the hysteresis effect, the dip in the graph when the photoresistor is blocked is slightly asymmetric. To avoid these problems, a double flag method is used where the time from the beginning of two separate dips is used for timing purposes. In this method, two blockages of the gate are used to make a single velocity measurement. This can be done using a piece of Plexiglas with two tape strips or a card with a notch cut-out, placed on the moving object so that two dips are produced at each photoresistor when the object passes and blocks the light to each resistor. The Android app AudioTime+ marks the gate as blocked at the time when the amplitude of the gradual signal drop is 80% of the maximum, unblocked signal. The time interval from the 80% mark at the first dip to the 80% mark at the second dip is the length of time the object takes to pass the photoresistor. K. Forinash et al. [65] used two photoresistors and a toy car with a notched card attached to measure acceleration down an inclined plane. For an inclination of 9.5 degrees, the acceleration of a frictionless sliding object would be g sin (θ) = 1.6 m/s^2. The notch on the car was cut out so that the leading edge of the two flags was 9.0 cm. A consistent acceleration of 1.3 m/s^2 over several trials was measured with the AudioTime+ phone app acting as a timer. The error is attributable to friction and rotational inertia in the wheels of the toy car. We also measured the acceleration of gravity using a free-falling Plexiglas square with two-tape strips, 11.9 cm apart. An average acceleration of 9.9 m/s^2 was measured over five trials. Photogates are very portable, inexpensive, and can be used in many different settings. For example, a single photoresistor can be used to measure the rotational motion of a wheel. This can be done by placing a small card with a known width on the wheel and arranging the photogate to be blocked by the card as the wheel rotates. Both constant angular speed and angular acceleration can be measured from the length of time of the amplitude dips in the output signal. Due to the portability and simplicity of the measurement apparatus, this works for a mounted wheel in a science lab or the wheel of a moving object such as a student's bicycle. As a final example, a photoresistor/phone combination attached to an object (bicycle, car, runner) moving past a series of openings of known spacing (openings in a picket fence, telephone poles, etc.) could be used to measure speed and acceleration. Because the apparatus is mobile, the exploration of the movement of many objects outside the ordinary classroom laboratory becomes possible. Table 6.1 mentions application of various photoresistors in a tabulated form.

TABLE 6.1
Variety of Photoresistor with Specified Application

Material Used	Source of Light	Operating Temp. (K)	Intensity	Resistance Variation	Application	Ref
Si-ZnO	UV light	300	(20 μW)	50% changed	Detector	[66]
Boron-doped SWCNT	–	300	–	613 (MΩ)	Light-dependent electronic devices.	[41]
Cd sulphide	X-ray	300	5 mW/cm^2	130 Ω changed	X-ray dosimetry	[67]
BNL-1 silicon with graphite	IR light-emitting diode (LED)	300	–	2.0–4.15 kΩ	Multifunctional optoelectronic applications	[68]
GaN	Xe arc lamp	300–680	–	–	Reduction in the bandwidth of the device	[2]
Silica nanopillars with CdS	UV light	300	400–11,000 μW/cm^2	Decreases rapidly	Used in higher photosensitivity response device	[69]
MoS$_2$ nanosphere	White light	473	–	–	Photo-resistive switching	[70]
ZnO films	UV light	300	1.3 mW/cm^2	4×10^{10}–1.7×10^5 Ω	Ultraviolet photoresistors	[71]
PbI$_2$	UV-visible light	373	7.8 A/W	–	High-performance photoresistors	[72]

6.6 CONCLUSION

The photoresistor is an electronic device that is sensitive to the light incident upon it. The light produces the change in the resistance of the photoresistor and this can be used for various application purposes. Mostly, a highly resistive and sensitive semiconductor material is used for fabricating a photoresistor. When the light is incident on such a material, the valence electrons move from the valence to the conduction band of the material. This results in a change in the resistance of the material. As a larger number of electrons move towards the conduction band, a large variation will be observed in the resistance of the material. The photoresistor is usually made of semiconductor materials such as PbS, PbSe, CdS, CdSe, and ZnO. Since at nanoscale, the properties of a material change drastically as compared to the bulk of the same material, the nanomaterials can be used for fabricating photoresistors. Progress has already been made in this regard as CNTs and nanowires of ZnO and some other materials have been successfully prepared with resistance dependent upon the intensity of light. The photoresistors have already been used for measuring the intensity of light, as optical biosensors, in street lighting, etc. The variety of materials that can be used as photoresistors is not very promising as only certain kinds of materials are used for photoresistor applications. The research needs to be done in searching for various new and more efficient materials that can be responsive to light. Moreover, the application of these photoresistors needs to be increased to various other fields of research and technology. The future is bright for photoresistors because they can prove to be very effective in meeting the energy requirements of the people due to their cost-effectiveness, easy fabrication, and high sensitivity towards light.

REFERENCES

1. Anekal, B., 2018. IOT based fault diagnostic device for photovoltaic panels, IJERECE.
2. De Vittorio, M., Potì, B., Todaro, M.T., Frassanito, M.C., Pomarico, A., Passaseo, A., Lomascolo, M. and Cingolani, R., 2004. High temperature characterization of GaN-based photodetectors. Sensors and Actuators A: Physical, 113(3), pp.329–333.
3. Hur, S.G., Kim, E.T., Lee, J.H., Kim, G.H. and Yoon, S.G., 2008. Enhancement of photosensitivity in CdS thin films incorporated by hydrogen. Electrochemical and Solid State Letters, 11(7), p.H176.
4. Zhang, K., Hu, C., Tian, Y., Zheng, C. and Wan, B., 2011. Stable and highly photosensitive device of CdSe nanorods. Physica E: Low-dimensional Systems and Nanostructures, 43(4), pp.943–947.
5. Pentia, E., Pintilie, L., Matei, I., Botila, T. and Pintilie, I., 2003. Combined chemical–physical methods for enhancing IR photoconductive properties of PbS thin films. Infrared Physics & Technology, 44(3), pp.207–211.
6. Simma, M., Lugovyy, D., Fromherz, T., Raab, A., Springholz, G. and Bauer, G., 2006. Deformation potentials and photo-response of strained PbSe quantum wells and quantum dots. Physica E: Low-Dimensional Systems and Nanostructures, 32(1–2), pp.123–126.
7. Liu, F., Zhang, R., Hu, Z., Sun, J., Huang, H., Li, Z., Zhao, J., Yin, P., Guo, L., Zhang, X. and Wang, Y., 2010. Structural and optical properties of nonpolar $(11\bar{2}0)$ ZnO films grown by plasma-assisted molecular-beam epitaxy. IEEE Transactions on Plasma Science, 39(2), pp.700–703.
8. Melnick, D.A., 1957. Zinc oxide photoconduction, an oxygen adsorption process. The Journal of Chemical Physics, 26(5), pp.1136–1146.
9. Witkowski, B.S., Wachnicki, L., Gieraltowska, S., Sybilski, P., Kopalko, K., Stachowicz, M. and Godlewski, M., 2014. UV detector based on zinc oxide nanorods obtained by the hydrothermal method. Physica Status Solidi (c), 11(9–10), pp.1447–1451.
10. Waser, R., Dittmann, R., Staikov, G. and Szot, K., 2009. Redox-based resistive switching memories–nanoionic mechanisms, prospects, and challenges. Advanced Materials, 21(25–26), pp.2632–2663.
11. Kapitanova, O.O., Panin, G.N., Kononenko, O.V., Baranov, A.N. and Kang, T.W., 2014. Resistive switching in graphene/graphene oxide/ZnO heterostructures. Journal of the Korean Physical Society, 64(10), pp.1399–1402.
12. Zhuge, F., Dai, W., He, C.L., Wang, A.Y., Liu, Y.W., Li, M., Wu, Y.H., Cui, P. and Li, R.W., 2010. Nonvolatile resistive switching memory based on amorphous carbon. Applied Physics Letters, 96(16), p.163505.

13. Qi, J., Olmedo, M., Ren, J., Zhan, N., Zhao, J., Zheng, J.G. and Liu, J., 2012. Resistive switching in single epitaxial ZnO nanoislands. ACS Nano, 6(2), pp.1051–1058.

14. Pantel, D., Goetze, S., Hesse, D. and Alexe, M., 2011. Room-temperature ferroelectric resistive switching in ultrathin Pb (Zr0.2Ti0.8) O3 films. ACS Nano, 5(7), pp.6032–6038.

15. Hu, B., Zhuge, F., Zhu, X., Peng, S., Chen, X., Pan, L., Yan, Q. and Li, R.W., 2012. Nonvolatile bistable resistive switching in a new polyimide bearing 9-phenyl-9H-carbazole pendant. Journal of Materials Chemistry, 22(2), pp.520–526.

16. Bessonov, A.A., Kirikova, M.N., Petukhov, D.I., Allen, M., Ryhänen, T. and Bailey, M.J., 2015. Layered memristive and memcapacitive switches for printable electronics. Nature Materials, 14(2), pp.199–204.

17. Maksymovych, P. et al., 2009. Polarization control of electron tunneling into ferroelectric surfaces. Science, 324, pp.1421–1425.

18. Polyakov, B., Daly, B., Prikulis, J., Lisauskas, V., Vengalis, B., Morris, M.A., Holmes, J.D. and Erts, D., 2006. High-density arrays of germanium nanowire photoresistors. Advanced Materials, 18(14), pp.1812–1816.

19. Tománek, D., Jorio, A., Dresselhaus, M.S. and Dresselhaus, G., 2007. Introduction to the important and exciting aspects of carbon-nanotube science and technology. In: Jorio A., Dresselhaus G., Dresselhaus M.S. (eds), Carbon Nanotubes. Topics in Applied Physics, vol 111. Springer, Berlin, Heidelberg. https://doi.org/10.1007/978-3-540-72865-8_1.

20. Avouris, P. and Martel, R., 2010. Progress in carbon nanotube electronics and photonics. MRS Bulletin, 35(4), pp.306–313.

21. Shah, K.A. and Parvaiz, M.S., 2016. Computational comparative study of substitutional, endo and exo BN Co-Doped single walled carbon nanotube system. Superlattices and Microstructures, 93, pp.234–241.

22. Wang, S., Zhang, Z. and Peng, L., 2012. Doping-free carbon nanotube optoelectronic devices. Chinese Science Bulletin, 57(2), pp.149–156.

23. Ueda, A., Matsuda, K., Tayagaki, T. and Kanemitsu, Y., 2008. Carrier multiplication in carbon nanotubes studied by femtosecond pump-probe spectroscopy. Applied Physics Letters, 92(23), p.233105.

24. Mutlu, M.U., Akın, O. and Yildiz, ÜH., 2018 Polymer nanofiber-carbon nanotube network generating circuits. Organic Photonic Materials and Devices, 10529, p. 105290R). International Society for Optics and Photonics.

25. Sze, S.M., 1981. Physics of Semiconductor Devices, J. Wiley & Sons. M. Sze Modern Semiconductor Device Physics. Wiley & Sons K. Seeger Semiconductor.

26. Piprek, J., 2013. Semiconductor Optoelectronic Devices: Introduction to Physics and Simulation. Elsevier.

27. Lutz, G., 2007. Semiconductor Radiation Detectors (pp. 978–3540716785). Heidelberg: Springer.

28. Deleplanque, M.A., Lee, I.Y., Vetter, K., Schmid, G.J., Stephens, F.S., Clark, R.M., Diamond, R.M., Fallon, P. and Macchiavelli, A.O., 1999. GRETA: utilizing new concepts in γ-ray detection. Nuclear Instruments and Methods in Physics Research Section A: Accelerators, Spectrometers, Detectors and Associated Equipment, 430(2-3), pp.292–310.

29. Rieke, G. and George, R., 2003. Detection of Light: From the Ultraviolet to the Submillimeter. Cambridge University Press.

30. Ranger, N.T., 1999. The AAPM/RSNA physics tutorial for residents: radiation detectors in nuclear medicine. Radiographics, 19(2), pp.481–502.

31. Lauhon, L.J., Gudiksen, M.S., Wang, D. and Lieber, C.M., 2002. Epitaxial core–shell and core–multishell nanowire heterostructures. Nature, 420(6911), pp.57–61.

32. Fujiki, M., Koe, J.R., Terao, K., Sato, T., Teramoto, A. and Watanabe, J., 2003. Optically active polysilanes. Ten years of progress and new polymer twist for nanoscience and nanotechnology. Polymer Journal, 35(4), pp.297–344.

33. Banerjee, S., Dan, A. and Chakravorty, D., 2002. Review synthesis of conducting nanowires. Journal of Materials Science, 37(20), pp.4261–4271.

34. Rabin, O., Herz, P.R., Lin, Y.M., Akinwande, A.I., Cronin, S.B. and Dresselhaus, M.S., 2003. Formation of thick porous anodic alumina films and nanowire arrays on silicon wafers and glass. Advanced Functional Materials, 13(8), pp.631–638.

35. Hamberg, I. and Granqvist, C.G., 1986. Optical properties of transparent and heat-reflecting indium tin oxide films: refinements of a model for ionized impurity scattering. Journal of Applied Physics, 59(8), pp.2950–2952.

36. Witkowski, B.S., Wachnicki, L., Giera-towska, S., Dluzewski, P., Szczepanska, A., Kaszewski, J. and Godlewski, M., 2014. Ultra-fast growth of the monocrystalline zinc oxide nanorods from the aqueous solution. International Journal of Nanotechnology, 11(9-1011), pp.758–772.

37. Huang, C.Y., Yang, Y.J., Chen, J.Y., Wang, C.H., Chen, Y.F., Hong, L.S., Liu, C.S. and Wu, C.Y., 2010. p-Si nanowires/SiO 2/n-ZnO heterojunction photodiodes. Applied Physics Letters, 97(1), p.013503.

38. Zhang, Z., Liao, Q., Yu, Y., Wang, X., and Zhang, Y., 2014. Enhanced photoresponse of ZnO nanorods-based self-powered photodetector by piezotronic interface engineering. Nano Energy, 9, pp.237–244.

39. Barkelid, M. and Zwiller, V., 2014. Photocurrent generation in semiconducting and metallic carbon nanotubes. Nature Photonics, 8(1), pp.47–51.

40. St-Antoine, B.C., Ménard, D. and Martel, R., 2009. Position sensitive photothermoelectric effect in suspended single-walled carbon nanotube films. Nano Letters, 9(10), pp.3503–3508.

41. Parvaiz, M.S., Shah, K.A., Dar, G.N. and Misra, P., 2020. Computational modeling of carbon nanotubes for photoresistor applications. Solid State Communications, 309, p.113831.

42. Li, B. and Li, Z., 2017, April. Design and implementation of a simple acousto optic dual control circuit. In AIP Conference Proceedings (Vol. 1834, No. 1, p. 020016). AIP Publishing LLC.

43. Tsybrii-Ivasiv, Z.F., Darchuk-Korovina, L.O., Sizov, F.F., Golenkov, O.G., Bilevych, Y.O., Sidorov, Y.G. and Varavin, V.S., 2004. Investigation and characterization of $Hg_{1-x}Cd_xTe$ epilayers. Journal of Alloys and Compounds, 382(1-2), pp.288–291.

44. Lin, C.H. and Chen, C.H., 2008. Sensitivity enhancement of capacitive-type photoresistor-based humidity sensors using deliquescent salt diffusion method. Sensors and Actuators B: Chemical, 129(2), pp.531–537.

45. Lodeiro, C., Capelo, J.L., Mejuto, J.C., Oliveira, E., Santos, H.M., Pedras, B. and Nuñez, C., 2010. Light and colour as analytical detection tools: a journey into the periodic table using polyamines to bio-inspired systems as chemosensors. Chemical Society Reviews, 39(8), pp.2948–2976.

46. Velasco-Garcia, M.N., 2009, February. Optical biosensors for probing at the cellular level: A review of recent progress and future prospects. In Seminars in Cell & Developmental Biology (Vol. 20, No. 1, pp. 27–33). Academic Press.

47. Newman, J.D. and Turner, A.P., 2005. Home blood glucose biosensors: a commercial perspective. Biosensors and Bioelectronics, 20(12), pp.2435–2453.

48. Moreno-Bondi, M.C., Wolfbeis, O.S., Leiner, M.J. and Schaffar, B.P., 1990. Oxygen optrode for use in a fiber-optic glucose biosensor. Analytical Chemistry, 62(21), pp.2377–2380.

49. Zhuo, D.H., Rai, A., Vosoogh-Grayli, S., Leach, G.W. and Bahreyni, B., 2020. A micromachined vector light sensor. Sensors and Actuators A: Physical, 311, p.112045.

50. Dutta, S., Mandal, N. and Bandyopadhyay, D., 2016. Paper based α-amylase detector for point-of-care diagnostics. Biosensors and Bioelectronics, 78, pp.447–453.

51. Agarwal, D. and Aswani, R., 2015. Microcontroller Based Light Control System.

52. Thom, N.K., Yeung, K., Pillion, M.B. and Phillips, S.T., 2012. "Fluidic batteries" as low-cost sources of power in paper-based microfluidic devices. Lab on a Chip, 12(10), pp.1768–1770.

53. Wang, L., 2015, April. Characteristics Test of Photoresistor and Its Application in Optical Control Switch. In International Conference on Advances in Mechanical Engineering and Industrial Informatics (AMEII 2015-26) Data sheet of the photo-light resistor.

54. Wong, E.Y., Zhang, Q., Duerk, J.L., Lewin, J.S. and Wendt, M., 2000. An optical system for wireless detuning of parallel resonant circuits. Journal of Magnetic Resonance Imaging: An Official Journal of the International Society for Magnetic Resonance in Medicine, 12(4), pp.632–638.

55. Saravanan, C., Panneerselvam, M.A. and Christopher, I.W., 2011. A novel low cost automatic solar tracking system. International Journal of Computer Applications, 31(9), pp.62–67.

56. Subramanyam, B.K., Reddy, K.B. and Reddy, P.A.K., 2013. Design and development of intelligent wireless street light control and monitoring system along with gui. International Journal of Engineering Research and Applications (IJERA), 3(4), pp.2115–2119.

57. Wang, L., 2015, April. Characteristics Test of Photoresistor and Its Application in Optical Control Switch. In International Conference on Advances in Mechanical Engineering and Industrial Informatics (AMEII 2015)[26] Data sheet of the photo-light resistor.

58. Byers, J.A. and Unkrich, M.A., 1983. Electronic light intensity control to simulate dusk and dawn conditions. Annals of the Entomological Society of America, 76(3), pp.556–558.

59. Nau, S., Wolf, C., Sax, S. and List-Kratochvil, E.J., 2015. Organic non-volatile resistive photo-switches for flexible image detector arrays. Advanced Materials, 27(6), pp.1048–1052.

60. Pierre, A., Gaikwad, A. and Arias, A.C., 2017. Charge-integrating organic heterojunction phototransistors for wide-dynamic-range image sensors. Nature Photonics, 11(3), pp.193–199.

61. Yang, X., Yu, Y., Shirowzhan, S. and Li, H., 2020. Automated PPE-tool pair check system for construction safety using smart IoT. Journal of Building Engineering, 32, p.101721.

62. Theuwissen, A.J., 2008. CMOS image sensors: state-of-the-art. Solid-State Electronics, 52(9), pp.1401–1406.

63. Chen, Q., Lin, M., Fang, Y., Wang, Z., Yang, Y., Xu, J., Cai, Y. and Huang, R., 2018. Integration of bio-compatible organic resistive memory and photoresistor for wearable image sensing application. Science China Information Sciences, 61(6), pp.1–8.

64. Forinash, K. and Wisman, R.F., 2012. Smartphones–experiments with an external thermistor circuit. The Physics Teacher, 50(9), pp.566–567.

65. Forinash, K. and Wisman, R.F., 2015. Photogate timing with a smartphone. The Physics Teacher, 53(4), pp.234–235.

66. Witkowski, B.S., Pietruszka, R., Gieraltowska, S., Wachnicki, L., Przybylinska, H. and Godlewski, M., 2017. Photoresistor based on ZnO nanorods grown on a p-type silicon substrate. Opto-Electronics Review, 25(1), pp.15–18.

67. Odeh, A.S., 1994. X-ray dosimetry using a Cd sulphide photoresistor. Radiation Physics and Chemistry, 44(1–2), pp.61–62.

68. Denisov, B.N., 2007. A photoresistor as a multifunctional optoelectronic element. Journal of Communications Technology and Electronics, 52(4), pp.478–481.

69. Liu, J., Liang, Y., Wang, L., Wang, B., Zhang, T. and Yi, F., 2016. Fabrication and photosensitivity of CdS photoresistor on silica nanopillars substrate. Materials Science in Semiconductor Processing, 56, pp.217–221.

70. Wang, W., Panin, G.N., Fu, X., Zhang, L., Ilanchezhiyan, P., Pelenovich, V.O., Fu, D. and Kang, T.W., 2016. MoS_2 memristor with photoresistive switching. Scientific Reports, 6(1), pp.1–11.

71. Liu, F.J., Hu, Z.F., Sun, J., Li, Z.J., Huang, H.Q., Zhao, J.W., Zhang, X.Q. and Wang, Y.S., 2012. Ultraviolet photoresistors based on ZnO thin films grown by P-MBE. Solid-State Electronics, 68, pp.90–92.

72. Li, J., Li, H., Ding, D., Li, Z., Chen, F., Wang, Y., Liu, S., Yao, H., Liu, L. and Shi, Y., 2019. High-performance photoresistors based on perovskite thin film with a high PbI2 doping level. Nanomaterials, 9(4), p.505.

7 Semiconductor Photovoltaic

Jazib Ali, Fateh Ullah, Rizwan Haider, Ghulam Abbas Ashraf,
Fahmeeda Kausar, Hamaela Razaq, and Hafeez Anwar

CONTENTS

7.1 INTRODUCTION

Direct sunlight can be transformed to electric current without the interference of heat engine, which is the role of photovoltaic technologies. The design of photovoltaic devices is craggy and simple, and one of the huge assets being their manufacturing is a system based on a stand-alone formula in which outputs of microwatts can be easily converted into megawatts. However, these devices are applied for pumping of water, reverse plant osmosis, energy and megawatt-scale power plants, communication systems, space vehicles, satellites, remote structures, and solar home systems. The demand of these photovoltaic devices is enhancing every year due to such a vast variety of applications.

7.2 GENERATION OF PHOTOVOLTAIC ENERGY

Global demand for energy at present is about 17.7 TW (Adm. 2017), and this value is expected to increase to 30 TW in 2050 (NASA 2020). Although the generation of energy from the burning of fossil fuels is an easy way, but its impact on the planet is very severe in the form of global warming. Reported results and comparing them to the observed temperature variations are shown in Figure 7.1, which shows the anomaly of global temperature rise from 1880 to 2020, possibly due to the increase of carbon dioxide (CO_2) emission every year.

Over the years, academics and industries have worked together to find the answer of constant energy demand. Currently, the modern sources of energy employed in daily life are the outcome of years of research and development. The current challenges are not only a shift from the existing

DOI: 10.1201/9781003188582-7

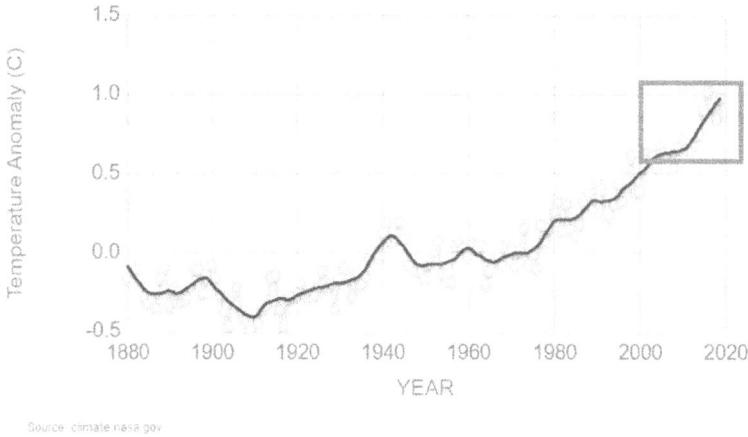

FIGURE 7.1   Global temperature anomaly from 1880 to 2020 (NASA 2020).

energy sources to cleaner ones but also a need to design and install capacities that meet up the future energy requirements. The progress of renewable energy sources is expected to satisfy both the challenges. Energy production using renewable sources is currently growing faster than all other means of energy production.

Among all the existed energy sources, solar energy attracts more interest. The conversion of sunlight into electric energy is a fundamental concept behind solar devices. Nonetheless, the term light here refers to the electromagnetic (EM) radiation radiated from the sun. These EM radiations cover the spectrum of frequencies similar to black body radiation at about 577K (Hegedus and Luque 2003), but only a small part of these radiations reach Earth's surface. For this purpose, an air mass coefficient (AM 1.5) was introduced to calculate the active EM radiations emitted by the sun at the sea level, which represents the maximum available energy converted by a solar cell into electrical energy (Wurfel 1982). According to the AM 1.5 spectrum, the highest irradiation peak observed is in the visible light region. The main loss in the original emission is due to the Earth's atmosphere contents such as absorption from oxygen, CO_2, ozone, and water vapors that stop the ultraviolet (UV) and the infrared (IR) part of the solar spectrum (Figure 7.2). An ideal solar cell can convert the entire spectral range to increase the yield of photo-generated current.

The photovoltaic (PV) technologies are categorized into three generations, subject to the nature of applied materials and the level of the commercial maturity of the technology. Traditional solar cells (first generation) composed of the silicon (Si) wafer (mono and polycrystalline) technology. The power conversion efficiency (PCE) of these monocrystalline Si devices reached to 25.6%, whereas 20.8% was for the polycrystalline Si (Polman et al. 2016). The Si solar cell's (wafer technology) efficiency in commercial modules was about 20%, with more than 20 years of lifetime, and it contributed about 93% of the total world PV market (Green et al. 2017). The major disadvantage of the Si solar cell was its expensive material and the high-temperature fabrication which increases the cost of these devices (Di Giacomo et al. 2016).

The second-generation solar cells were based on "thin film" technology such as single crystalline gallium arsenide (GaAs-28.8%), polycrystalline cadmium telluride (CdTe-21.5%), and copper indium gallium diselenide (CIGS-21.7%) (Polman et al. 2016). The high cost of these solar devices is still a major challenge for their commercialization. Researchers introduced a "third-generation" technology to overcome the drawback of the previous PV

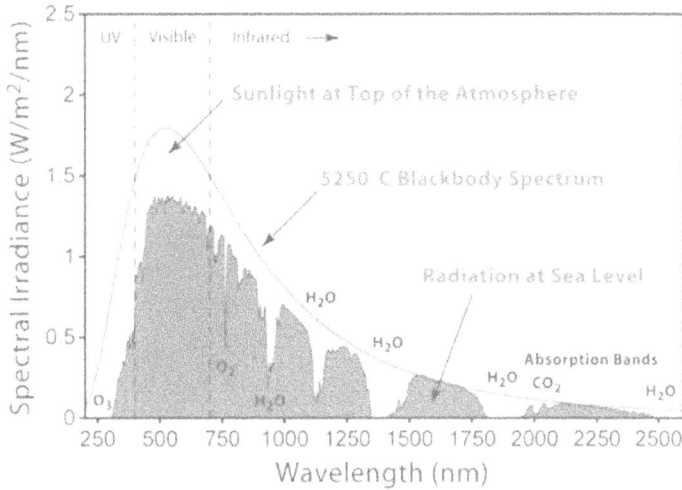

FIGURE 7.2 Solar irradiance spectrum above the atmosphere and at the surface (NREL 2021).

technologies, which includes the dye-sensitized solar cells (DSSCs), organic solar cells (OSCs), etc. (Rinaldi 2013). All of these PV technologies are used as low-cost fabrication and exhibited promising PCEs.

In this chapter, we have summarized different semiconducting PV technologies or classes that are under research phase or commercialization phase.

7.3 SILICON-CRYSTALLINE STRUCTURE

The crystalline Si is termed as the 1st generation of solar cells to construct the PV architectures, which are then joined to manufacture the PV modules. Nonetheless, such technologies were modified constantly in order to accomplish an enhanced performance and device efficiencies. Different systems including the mono and polycrystalline all together direct to the roof of crystalline Si and are reviewed in the succeeding sections.

7.4 MONOCRYSTALLINE PHOTOVOLTAIC DEVICES

The mono-crystalline PV technology is most frequently implemented, occupied an approximately 80% of the world market, and will competitively lead until the development of alternative, efficient, and cost-effective PV candidate. The monocrystalline PV fundamentally employs crystalline Si p-n junctions, and the Czochralski technique is applied for exploiting a single crystal ingot for attaining a monocrystalline silicon (Wikipedia 2020). The PCE delivered by monocrystalline Si-PV is limited by the characteristics features of silicon material, because the energy generated by photons shrinks as higher wavelength are approached. Additionally, the longer wavelengths radiation governs thermal dissipation, eventually condenses the efficiency attributed to the heating up of the cell. Although the maximum efficiency for monocrystalline silicon PV cell has been stretched around 23% under standard test condition (STC), the highest recorded was 24.7% (under STC).

Furthermore, due to the resistance produced by the solar cell coupled with reflection of solar radiations and corresponding top metal contacts, self-losses are generated. Once the manufactured Si ingot reached to diameter under 10–15 cm, it is followed by tailoring into wafers with 0.3-mm thickness, eventually forming a solar device producing ≈0.55 V voltage and 35 m/A^2 of current density at full illumination. Considering other semiconducting materials with varied wavelengths, it can reach 30% (under STC). However module efficiencies always tend to be lower than the actual cell and Sunpower (Sunpower 2021). Recent measurements done by national renewable energy laboratory (NREL) claimed a 20.4% full panel efficiency revealing the best ever efficiency (NREL 2021), while providing enduring stability and demonstrating good compatibility with the prevailing market sources. Technologically, the silicon-based solar processing industries and the microelectronic industries are sharing common values that are beneficial for the enormous developments in the processing technologies of Si wafer to implement in microelectronic applications which helped to increase the overall PCE of these solar cells.

7.5   POLYCRYSTALLINE PHOTOVOLTAIC PANELS

Innovative crystallization procedures have been employed in accomplishing simultaneously the production costs and throughputs. Initially, the polycrystalline silicon PVs were dominating the solar industry, despite the fact that the cost of Si was $340/kg. However, the value of silicon reduced to $50/kg, such technology presents an efficiency of 15% that is lower in comparison to the monocrystalline silicon, but lower manufacturing cost is more attractive feature of this technology. The benefit of transforming the manufacturing of crystalline solar cells from mono to multi-silicon is to reduce the defects in crystal structure (Manna and Mahajan 2007). Manufacturing process of polycrystalline solar cell begins with the melting of silicon followed by solidification to attain orient crystals while succeeding directionally, constructing rectangular ingot of multi-crystalline silicon. The Evergreen solar was first developed this technology (Solar 2021).

7.6   SILICON-CRYSTALLINE INVESTMENT

PV systems have huge starting capital expenses; however, little intermittent expenses for maintenance and operation. It is noteworthy that output energy prices decrease with an increase in lifetime of PV system and vice versa. A lifetime of 20–30 years has been reported for the aforementioned silicon-based systems. In most of the systems, longer restitution time provided heavy government incentives. Hence, it is need of the day to make PVs a practical innovation that can remain all alone without hefty government sponsorships. To achieve that goal, many researchers are focusing on to reduce the initial capital cost resulting in reduced restitution time. Thin-film solar cells (T-FSCs) are in spotlight because of the need to decrease the assembling, and hence module cost. A definitive objective is actuality the accomplishment of "grid purity" that makes the output expense of the kWh by PV systems comparable to the kW h by conventional technologies. Although technology improvement has reduced per watt cost considerably, but the above goal stays elusive till today.

7.7   THIN-FILM TECHNOLOGY

The thin-film technology is a promising alternative to crystalline silicon cells, by potentially reducing the cost of a PV array in terms of manufacturing of materials and synthesis process devoid of endangering the life span of device as well as initiating damage to environment. By exploiting the sputtering apparatuses, the deposition of certain thin-layered materials like glass or stainless steel to form thin-film panels and to obtain solar modules, glass panels unlike crystalline solar cells are engaged to sandwich the semiconductor materials. A very advantageous methodology is the deposition of thin films on sheets of stainless steel, which allows the flexible PV module's manufacturing; additionally, the crystalline wafers are several hundred microns thick and deposited layers are up to few microns (smaller than 10 m) thick. Technically, the individual active layers are so thin that the PV material absorbs limited solar radiation, so the energy conversion efficiency of thin-film solar modules is less than that of crystalline material. This technology has the potential to accumulate many different materials and, although it is a composite, makes an excellent improvement in the efficiency of the solar cell. Besides, the versatility and temperature stability of the thin-film photovoltaic modules increased their contribution to the market by 15–20% as shown in Figure 7.3.

The architecture of T-FSCs fundamentally incorporates semiconductor's thin film that works as a solid backing material. This kind of photovoltaic cells significantly decrease the quantity of semiconducting material needed for individual cell in comparison to silicon wafers and produced cost-effective solar cells. Copper indium diselenide ($CuInSe_2$), GaAs, and CdTe are the fundamental materials implemented in thin-film PV cells. An investigation carried out by Barnett and coworkers revealed that incorporation of polycrystalline silicon into T-FSCs can potentially deliver the photovoltaic PCEs greater than 19%, attributed to the back surface elevation by optimum silicon thickness and light trapping (Barnett et al. 2001). Besides, studies of Albere suggested that auspicious thin-film c-Si PV technologies (developed through the last decade) and the three diverse technologies, i.e., SILVER, hybrid, CSG that are the thin film c-Si PV have the potential to be industrialized (Aberle 2006). Fave and coworkers conducted studies related to the comparison of growth of thin films of silicon epitaxial on the sacrificial layers that were doubly porous and attained via phase epitaxy, i.e., liquid and vapor (LPE or VPE). They observed that the length of dispersion and mobility was slightly higher with VPE compared to LPE during the fabrication of solar cells while exploiting an isolated film displays 4.2% efficiency rate having fill factor 0.69, that was obtained by VPE and irrespective of antireflective coating or surface passivation treatment (Fave et al. 2004). Reflection high energy electron diffraction (RHEED) thin-film pattern of HgCdTe and CdTe grown on Si via laser deposition technique was reported by Sagan et al. (2005).

A novel methodology of transfer of thin spongy silicon layers on top of alumina (a ceramic substrate) was documented by Solanki et al. (2002). Powalla and Dimmler (2001) described

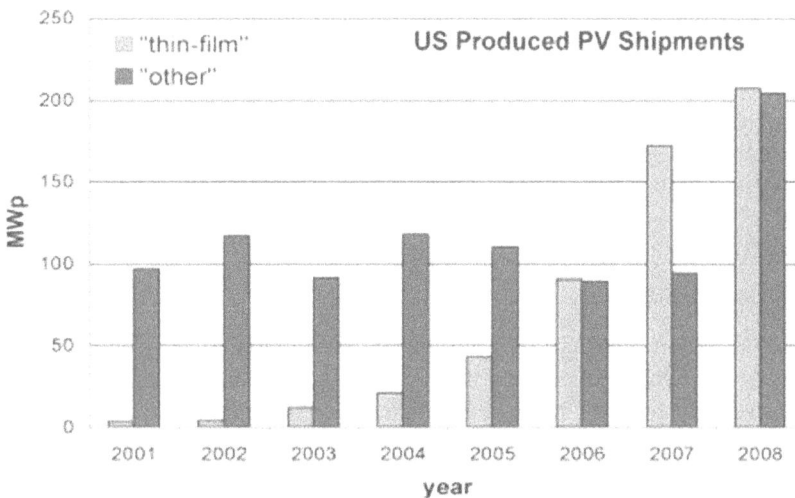

FIGURE 7.3 US-produced Modules through 2007, named thin film by counting First, Uni and Global Solar and remaining as "other." Projections for 2008 use 2/3 of PV News reported end of 2008 capacity figures. (Adapted with permission from von Roedern and Ullal (2008). Copyright 2008, 33rd IEEE Photovoltaic.)

that the reduction in manufacturing cost by keeping high fabrication volumes to project innovative challenge to associate appropriate yield with the high amount and higher quality was adopted by specially Cu(In,Ga)Se$_2$ (CIGS)-based thin-film technologies from all over technologies to attain more than 11% and a maximum of 12.7% efficiency. Furthermore, Hollingsworth et al. 2003 documented their studies on fabrication of thin films, unfolding that the deposition of CuGaS$_2$, CuGaInS$_2$, and CuInS$_2$ at lowest temperatures of 400–450° C by utilizing ternary single-source precursors either in hot or cold wall chemical vapor deposition reactor, displayed worthy optical and electrical properties appropriate for PV expedients. The fabrication of TiO$_2$ films was reported by Ito et al. (2008) for photosensitive solar cells comprising preliminary treatment of photoelectrode working through TiCl$_4$, differences in the size of TiO$_2$ nanocrystalline layer, and the use of a topcoat light diffusion coating and adhesion to the electrode transduction surface of global air weight 1.5 (AM 1.5, 1000 W/m^2) solar light at the energy efficiency of more than 10%.

Messina, Nair, and Nair (2007) proposed an approach for depositing thin films of Sb$_2$S$_3$ having thickness range of 500–600 nm from a single bath accompanying crystallinity enhancement. The thickness of the film, and photoconductivity onto annealing film in nitrogen, was favorable for their application in photovoltaic structures. Liehr and Dieguez-Campo (2005) presented that the higher rates of deposition required microwave plasma-enhanced chemical vapor deposition (PECVD) and for photovoltaic applications, enhanced shattering of precursors desired. The electric transport characteristics of zinc phthalocyanine thin films that were flash evaporated and method of direct current (DC) conduction of al-ZnPc-Al films at temperature variations as well as a potential mechanism of conduction in ZnPc films and field dependence behavior under DC field has been documented by Sathyamoorthy and coworkers (Sathyamoorthy et al. 2004).

7.8   AMORPHOUS SILICON

With the use of amorphous silicon in the fabrication of thin-film cells, cell efficiencies can be uplifted to 5–7% and even 8–10% in double and triple-junction designs, but they have a problem of degradation. Amorphous silicon nitride (a-SiN), amorphous silicon carbide (a-SiC), microcrystalline silicone (c-Si), and amorphous silicon germanium (a-SiGe) are some examples of amorphous silicon. Roll-to-roll continuous deposition was used by Yang, Banerjee, and Guha (2003) to pave the way for spectrum splitting triple-junction structure design resulting an air mass (AM) value of 1.5 and 13% stable cell efficiency. Nature of Staebler-Wronski effect was studied by Lund et al. (2001) in amorphous solar cells in the laboratory and field-on using different methods and operational conditions to reduce this effect. A series of fabrication technologies was developed by Tawada and Yamagishi (2001) for direct-super-straight-type modules using large area a-Si deposition technique with 8% yield.

7.9   CRYSTALLINE SILICON

In the case of crystalline silicon, cost-effective material is required which gives better performance in contrast to the amorphous silicon and multi-crystalline silicon solar cells (m-CSSCs) that are commercially available and have touched an efficiency of 14–19%. Green et al. (2004) combined the thin film and standard silicon wafer-based technology to develop solar cell technology of crystalline silicon on glass (CSG). It reduced the manufacturing cost to minimum and increased the efficiency from 8–9 to 12–13% for small pilot lone modules. It was reported by Shah et al. (2000) that low-temperature (200–250° C) deposition of intrinsic microcrystalline silicon using very high frequency-glow discharge (VHF-GD) method can be successfully used as photovoltaic active material for inverted and regular architecture type solar cells. Lipiipinski et al. (2002) improved the performance of standard screen printed solar cells by double porous silicon (d-PS) layers deposition on top surface of n+/p m-CSSCs by using acid chemical etching, which work as antireflection coating in the above system and boosted the efficiency to about 12%. The interaction between the workpiece and the laser was enhanced by a method of laser texturization for m-CSSCs (Dobrzaaski and Drygała 2007). The effect of emitter thickness on the photovoltaic properties of monocrystalline silicon solar cells with porous silicon was investigated by Vitanov et al. (2000). Electronically coupled up-conversion in crystalline silicon was reported by Macdonald et al., which displayed impurity-photovoltaic effect (IPV). As compared to the conventional IPV approach, it avoids two of the major problems named as recombination of minority charge carriers and parasitic absorption (Macdonald et al. 2007). Novel silver cells developed by Franklin et al. using single crystal silicon, which offered a 10–20-fold lesser silicon consumption. In addition, it required 20–40 times fewer wafer starts per MW as compared to conventional wafer-based technologies in an industrial production environment (Franklin et al. 2007).

7.10 CADMIUM TELLURIDE AND CADMIUM SULFIDE

CdTe/CdS solar cells fabricated by close spaced sublimation (CSS) technique were reported by Ferekides and coworkers. This system has attractive features for the production of large area solar cells for various applications along with high deposition rates and efficient material utilization (Ferekides et al. 2000). Additionally, Pfisterer explained the effect of surface treatments of the Cu_2S-CdS solar cells by the introduction of metallic or semiconducting layers at the surfaces having monolayer-range thicknesses. They also examined the influence of lattice mismatch on epitaxy as well as wet and dry topotaxy and preconditions for successful application of topotaxy (Pfisterer 2003). By the use of ray-tracing simulations, Richards and McIntosh (2007) demonstrated that the short-wavelength region of CdTe/CdS solar modules can be enhanced, but they achieved poor internal quantum efficiency on application of a luminescent downshifting layer to the solar module.

7.11 ORGANIC-SOLAR CELL TECHNOLOGY

The OSCs are another emerging photovoltaic technologies that are gaining popularity as promising contenders of silicon-based methodologies to address the current energy insecurities. Certain fundamental features such as mechanical flexibility, cost effectiveness and batch-to-batch material production, disposable, lightweight, and semitransparent nature of organic photovoltaics (OPV) are suitable for the future energy demands. In addition, to composed of molecular or polymeric organics (typically fabricating 100 nm of organic films, including polymers, small molecules, polyphenylene vinylene, pentacene, copper phthalocyanine, and fullerene and non-fullerene derivatives) (Piradi et al. 2019; Xue et al. 2019; Zhu et al. 2019) known as plastic photovoltaic. OSCs have been fabricated using solution processing techniques such as spin coating, blade coting, and printing but not utilized vacuum processing method (An et al. 2019; Li et al. 2018; Zhu et al. 2020). Therefore, they are cost and time effective and exhibit minimal influence on the environment throughout the manufacturing and operations (Li et al. 2018).

The working mechanism of OSCs involves the electron-hole pair. When the donor/acceptor blends are illuminated, penetration of incident photons (light energy) from the solar spectrum stimulates the photoactive layer with the generation of electron-hole pairs that are transported and collected before the recombination process occurs. The appealing properties of OSCs like semitransparent nature and mechanical flexibility opening the door in as building materials, but the lower PCE and long-term stability are major obstacles for their large-scale applications. Lots of research work has already been conducted by the scientific community in enhancing the PCE by using bulk heterojunctions or addition of third component in the form of donor or acceptor (Li et al. 2021). The PCE reported in 1986 by Tang (1986) has been increased from 1% to over 18% (Liu et al. 2020) in last 15 years as indicated in Figure 7.4.

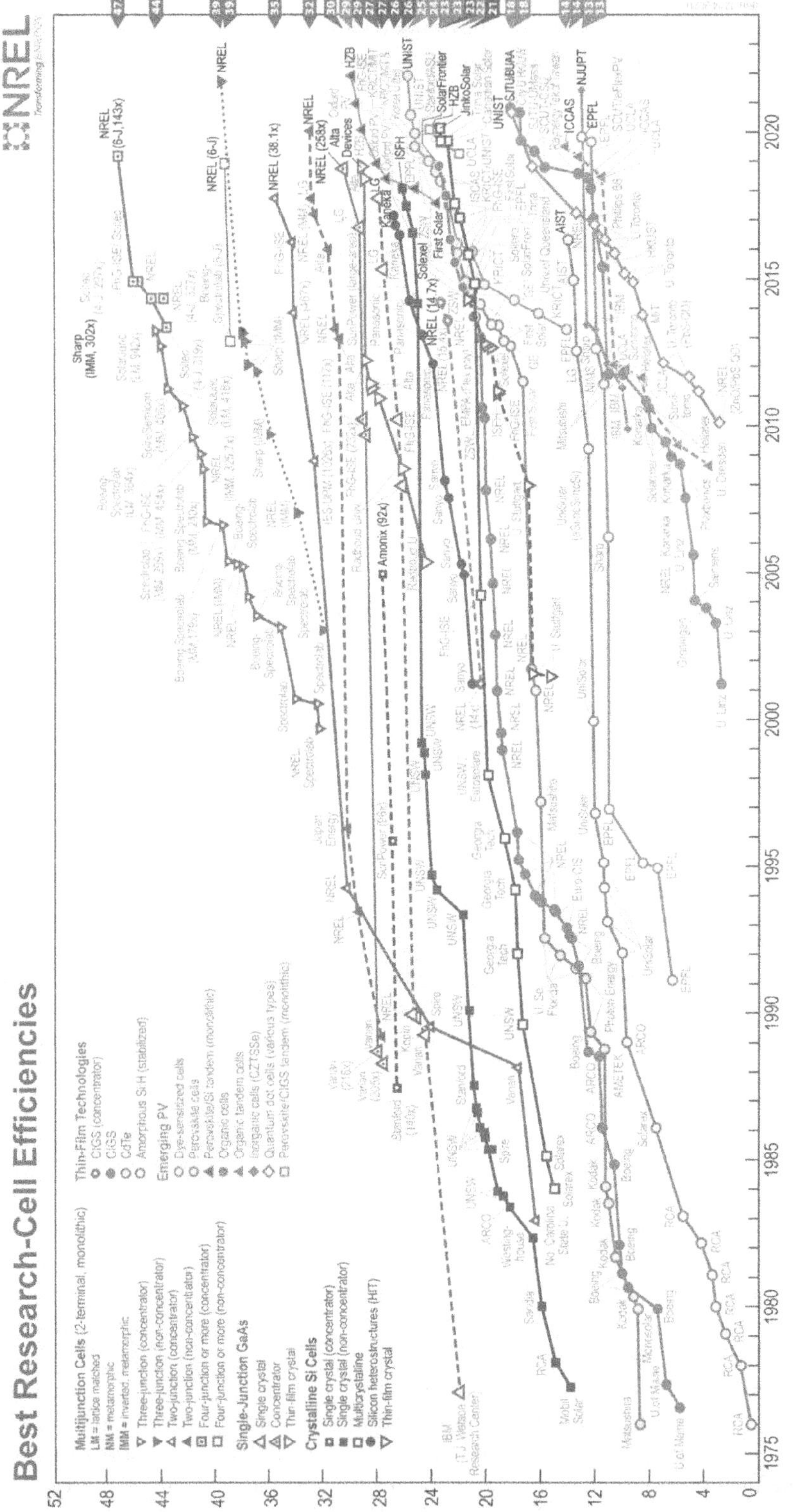

FIGURE 7.4 Best NREL solar cells efficiencies (NREL 2021).

7.12 DYE-SENSITIZED SOLAR CELL TECHNOLOGY

In recent scenario, the DSSC technology has gained significant attraction from researchers in academia and industry because of certain attractive features such as cost effectiveness, simple fabrication, exceptional, and appealing characteristics like colors and transparency. Besides, the DSSC technology addresses the environmental hazards associated with others semiconductor technologies (Akinyele, Rayudu, and Nair 2015; Ali et al. 2016; Pandey et al. 2016). The TiO_2 nanoparticles are deposited which are responsible for the attachment of dye molecules. The fabrication process includes the solution containing nanoscale TiO_2 particles placed with dye solution (left overnight) which eventually absorb the respective dye, and finally the dye molecules get anchored onto TiO_2 surface via chemical bonding, the procedure followed is known as sensitization (Andualem and Demiss 2018; Sugathan, John, and Sudhakar 2015).

On exposing DSSCs to the sunlight, the striking photon filled with light energy excites the dye molecules by delivering energy to the electrons, escaping electrons from the molecules followed by the transfer to TiO_2, which in turn left behind the holes on dye molecules. These electrons moved from counter electrode to the external circuit and generated power. However, lower PCE can be accomplished which is the prime concern of this technology. Besides, another problem associated to this technology is the introduction of electrolytes that encompassed volatile solvents, while working (DSSCs) at outdoor conditions solvents spread over the plastic and deteriorate solar cell encapsulation (Akinyele, Rayudu, and Nair 2015; Andualem and Demiss 2018). The scientific community is rigorously employing efforts for advancing the DSSCs technologies in achieving higher efficiencies. The recent studies revealed the PCE of DSSCs has approached to 13% (NREL 2021).

7.13 PEROVSKITE SOLAR CELL TECHNOLOGY

Recently, newly developed solution-processed perovskite solar cells (PSCs) has captured great importance due to their extraordinary performance and remarkable balance between cost and PCE (~25.5%) (Green 2003). This solar cell consists of a semiconductor material (perovskite mainly $CH_3NH_3PbI_3$) that works as an active layer and characterizes by its energy bandgap (an energy gap between the valence band [VB] and conduction band [CB] of semiconductor material). The bandgap determines the absorption properties of the semiconductor, during which only those photons having energy more than the bandgap of this material are absorbed, and then generate excitons. Coulomb interaction forces tightly bound these electrons and holes due to low permittivity in organic semiconductors. During the photovoltaic process, these excitons dissociated into free electron and hole and move through the active layer. Furthermore, these free carriers extracted to their relevant electron and hole transport layers and finally collected to the external circuit by an electrode, where it releases absorbed energy and generate power before returning to the opposite electrode and recombining with the hole by closing the cycle.

The PSCs have been in the focus since the last decade, because of their encouraging photovoltaic properties. In spite of the best-reported performance of mesoporous PSCs, the high temperature ≥500° C sintering restricting their use in the emerging flexible PSCs (Ali, Li, et al. 2020; Ali, Song, et al. 2019; Zhang et al. 2020). To conquer this issue, a planar inverted PSCs with a p-i-n structure was developed that requiring only low temperature (typically<150° C) solution-processed CTLs (Ali, Afreen, et al. 2019; Ali, Gao, et al. 2020).

7.14 HYBRID SOLAR CELL TECHNOLOGY

The combination of inorganic and organic semiconducting materials in conjunction forms a hybrid solar cell. Conventional solar architectures involved inorganic materials (largely silicon) which revealed elevated PCEs, but with high production cost. In contrast, the solar cells that mainly involved organic materials have low production costs and via molecular designs and modern synthetic tools, they can be functionalized along with ease of tuning their energy level (Pandey et al. 2016).

Therefore, hybrid solar cells perfectly blend the advantageous features of these two technologies and represent an emerging but vastly effective and inexpensive solar technology. Such amalgamation of solar technology is constructed of a heterojunction in combination with c-Si and a-Si fundamental thin layered tools.

A Japanese enterprise known as SANYO is one of the biggest production houses having 21% conversion efficiency, which produces hybrid PV solar cells synthetic intrinsic thin layer (HIT) modules (Akinyele, Rayudu, and Nair 2015). Commercially speaking, for similar module dimensions, the PCE of conventional c-Si modules is lower than that of HIT modules. Considering the identical size, SANYO 250W HIT module exhibited a PCE of 19%, whereas c-Si module by SUNTECH 250W comprises only 15.38%. Altogether, 19.4% and 15.1% efficiencies are accomplished by Panasonic 245 HIT module or SUNTECH 245W and Poly-Si module, respectively. Furthermore, an upcoming group of photovoltaic devices is the solution-processable thin-film photovoltaics, such as the hybrid organic-inorganic PSCs gathered significant attention due to their rapid increase of PCEs (Sheikh et al. 2015; Qiu et al. 2016). The top three electrodes, i.e., aluminum (Al), silver (Ag), and gold (Au), are mainly supplied by perovskite/SI tandems. The irregularities of inner transverse layer C of solar cells and P-N junctions were used to compare these three solar modules. When the variety of solar cells were compared, best energy return time and ecological outcomes were identified by the perovskite/Si Tandem that uses a top electrode, i.e., Al.

7.15 CONCLUSION

This chapter explains energy demand of the entire world, and its fulfillment is done by the burning of fossils fuels that produce trace amount of CO_2 and other gases. These hazardous gases badly affect the earth environment and rise its overall temperature. To overcome this situation, renewable energy sources must be used in place of these fossil fuels. So, photovoltaic cells have the ability to provide highest amount of clean and renewable energy as compared to any other source. Here, we thoroughly summarized different semiconductor materials (mono and polycrystalline solar cell, thin-film solar cell, amorphous silicon solar cell, organic solar cell, DSSC, PSC, and hybrid solar cell) that extensively used by the photovoltaic research community as well as industry.

REFERENCES

Aberle, Armin G. 2006. Fabrication and characterisation of crystalline silicon thin-film materials for solar cells. *Thin Solid Films* 511:26–34.

Adm., U.S. Energy Inf. *International Energy Outlook* 2017]. Available from www.eia.gov/outlooks/ieo/pdf/0484(2017)

Akinyele, Daniel, Rayudu, Ramesh, and Nair, Nirmal-Kumar C. 2015. Global progress in photovoltaic technologies and the scenario of development of solar panel plant and module performance estimation – Application in Nigeria. *Renewable and Sustainable Energy Reviews* 48:112–139.

Ali, Jazib, Afreen, Mutayyab, Rasheed, Tahir, Ashfaq, Muhammad Z., and Bilal, Muhammad. 2019. 18 applications of polyethylenedioxythiophene in photovoltaics. *Conducting Polymers-Based Energy Storage Materials.*

Ali, Jazib, Song, Jingnan, Li, Yu, et al. 2019. Control of aggregation and dissolution of small molecule hole transport layers via a doping strategy for highly efficient perovskite solar cells. *Journal of Materials Chemistry C* 7 (38):11932–11942.

Ali, Jazib, Gao, Peng, Zhou, Guanqing, et al. 2020. Elucidating the roles of hole transport layers in p-i-n perovskite solar cells. *Advanced Electronic Materials* 6 (12):2000149.

Ali, Jazib, Li, Yu, Gao, Peng, et al. 2020. Interfacial and structural modifications in perovskite solar cells. *Nanoscale* 12 (10):5719–5745.

Ali, N., Hussain, A., Ahmed, R., et al. 2016. Advances in nanostructured thin film materials for solar cell applications. *Renewable and Sustainable Energy Reviews* 59:726–737.

An, Yongkang, Liao, Xunfan, Chen, Lie, et al. 2019. A1-A2 type wide bandgap polymers for high-performance polymer solar cells: energy loss and morphology. *Solar RRL* 3 (1):1800291.

Andualem, Anteneh, and Demiss, Solomon. 2018. Review on dye-sensitized solar cells (DSSCs). *Edelweiss Applied Science and Technology* 2:145–150.

Barnett, Allen M, Rand, James A., Hall, Robert B., et al. 2001. High current, thin silicon-on-ceramic solar cell. *Solar Energy Materials and Solar Cells* 66 (1–4):45–50.

Di Giacomo, Francesco, Fakharuddin, Azhar, Jose, Rajan, and Brown, Thomas M. 2016. Progress, challenges and perspectives in flexible perovskite solar cells. *Energy & Environmental Science* 9 (10):3007–3035.

Dobrzanski, Leszek A., and Drygała, Aleksandra. 2007. Laser processing of multicrystalline silicon for texturization of solar cells. *Journal of Materials Processing Technology* 191 (1–3):228–231.

Fave, Alain, Quoizola, Sébastien, Kraiem, Jed, Kaminski, A., Lemiti, M., and Laugier, André. 2004. Comparative study of LPE and VPE silicon thin film on porous sacrificial layer. *Thin Solid Films* 451:308–311.

Ferekides, Christos S., Marinskiy, Dmitriy, Viswanathan, Vijay, et al. 2000. High efficiency CSS CdTe solar cells. *Thin Solid Films* 361:520–526.

Franklin, Evan, Everett, Vernie, Blakers, Andrew, and Weber, Klaus. 2007. Sliver solar cells: High-efficiency, low-cost PV technology. *Advances in Opto Electronics* 2007: 35383, 9.

Green, M.A. 2003. Impurity photovoltaic and multiband cells. *Third Generation Photovoltaics: Advanced Solar Energy Conversion book series (Photonics)* 12:95–109.

Green, Martin A., Basore, Paul A., Chang, Nathan., et al. 2004. Crystalline silicon on glass (CSG) thin-film solar cell modules. *Solar Energy* 77 (6):857–863.

Green, Martin A., Hishikawa, Yoshihiro, Warta, Wilhelm, et al. 2017. Solar cell efficiency tables (version 50). *Progress in Photovoltaics: Research and Applications* 25 (7):668–676.

Hegedus, Steven S., and Luque, Antonio. 2003. Status, trends, challenges and the bright future of solar electricity from photovoltaics. *Handbook of Photovoltaic Science and Engineering*. John Wiley & Sons, UK, Ltd, 1–43.

Hollingsworth, Jennifer A., Banger, Kulbinder K., Jin, Michael H.-C., et al. 2003. Single source precursors for fabrication of I–III–VI2 thin-film solar cells via spray CVD. *Thin Solid Films* 431:63–67.

Ito, Seigo, Murakami, Takurou N., Comte, Pascal, et al. 2008. Fabrication of thin film dye sensitized solar cells with solar to electric power conversion efficiency over 10%. *Thin Solid Films* 516 (14):4613–4619.

Li, Chao, Zhou, Jiadong, Song, Jiali, et al. 2021. Non-fullerene acceptors with branched side chains and improved molecular packing to exceed 18% efficiency in organic solar cells. *Nature Energy*.

Li, Zhenye, Xie, Ruihao, Zhong, Wenkai, et al. 2018. High-performance green solvent processed ternary blended all-polymer solar cells enabled by complementary absorption and improved morphology. *Solar RRL* 2 (10):1800196.

Liehr, Michael, and Dieguez-Campo, Manuel. 2005. Microwave PECVD for large area coating. *Surface and Coatings Technology* 200 (1–4):21–25.

Lipiace, Michał, Panek, Piotr, Świątek, Zbigniew, Bełtowska, Ewa, and Ciach, R. 2002. Double porous silicon layer on multi-crystalline Si for photovoltaic application. *Solar Energy Materials and Solar Cells* 72 (1–4):271–276.

Liu, Qishi, Jiang, Yufan, Jin, Ke., et al. 2020. 18% Efficiency organic solar cells. *Science Bulletin* 65 (4):272–275.

Lund, Chris P., Luczak, Kazimierz, Pryor, Trevor, et al. 2001. Field and laboratory studies of the stability of amorphous silicon solar cells and modules. *Renewable Energy* 22 (1–3):287–294.

Macdonald, Daniel, McLean, Kate, Deenapanray, Prakash N.K., De Wolf, Stefaan, and Schmidt, Jan. 2007. Electronically-coupled up-conversion: an alternative approach to impurity photovoltaics in crystalline silicon. *Semiconductor Science and Technology* 23 (1):015001.

Manna, Tapas K., and Mahajan, Swadesh M. 2007. Nanotechnology in the development of photovoltaic cells. *Paper read at 2007 International Conference on Clean Electrical Power.* DOI: 10.1109/ICCEP.2007.384240.

Messina, Sarah, Nair, Parasad, and Nair, Pankajakshy Karunakaran. 2007. Antimony sulfide thin films in chemically deposited thin film photovoltaic cells. *Thin Solid Films* 515 (15):5777–5782.

NASA. *Global Land-Ocean Temperature Index* 2020. Available from https://climate.nasa.gov/vital-signs/global-temperature/

NREL. *Best Research Solar-cell efficiency chart*. U.S. Department of Energy 2021. Available from https://www.nrel.gov/pv/cell-efficiency.html

NREL. *Reference Air Mass 1.5 Spectra* 2021. Available from https://www.nrel.gov/grid/solar-resource/spectra-am1.5.html

Pandey, Adarsh Kumar., Tyagi, Vineet, Jeyraj, Selvaraj, Rahim, Nasrudin Abd., and Tyagi, Sudhir. 2016. Recent advances in solar photovoltaic systems for emerging trends and advanced applications. *Renewable and Sustainable Energy Reviews* 53:859–884.

Pfisterer, Fritz. 2003. The wet-topotaxial process of junction formation and surface treatments of Cu_2StCdS thin-film solar cells. *Thin Solid Films* 431:470–476.

Piradi, Venkatesh, Xu, Xiaopeng, Wang, Zaiyu, et al. 2019. Panchromatic ternary organic solar cells with porphyrin dimers and absorption-complementary benzodithiophene-based small molecules. *ACS Applied Materials & Interfaces* 11 (6):6283–6291.

Polman, Albert, Knight, Mark, Garnett, Erik C., Ehrler, Bruno, and Sinke, Wim C. 2016. Photovoltaic materials: Present efficiencies and future challenges. *Science* 352 (6283):aad4424.

Powalla, M., and Dimmler, B. 2001. CIGS solar cells on the way to mass production: process statistics of a 30 cm × 30 cm module line. *Solar Energy Materials and Solar Cells* 67 (1-4):337–344.

Qiu, Jianhang, Wang, Gaoxiang, Xu, Wenjing, et al. 2016. Dark-blue mirror-like perovskite dense films for efficient organic–inorganic hybrid solar cells. *Journal of Materials Chemistry A* 4 (10):3689–3696.

Quanzeng, Zhang, Xiong, Shaobing, Ali, Jazib, Qian, Kun, Li, Yu, Feng, Wei, Hu, Hailin, Song, Jingnan, and Liu, Feng. 2020. Polymer interface engineering enabling high-performance perovskite solar cells with improved fill factors of over 82%. *Journal of Materials Chemistry C* 8 (16):5467–5475.

Richards, B. S., & McIntosh, K. R. (2007). Overcoming the poor short wavelength spectral response of CdS/CdTe photovoltaic modules via luminescence down-shifting: ray-tracing simulations. *Progress in Photovoltaics: Research and Applications* 15 (1):27–34.

Rinaldi, N. 2013. Solar PV module costs to fall to 36 cents per watt by 2017. *Greentech Media* 17. Available at: https://www.greentechmedia.com/articles/read/solar-pv-module-costs-to-fall-to-36-cents-per-watt

Sagan, Piotr, Wisz, Grzegorz, Bester, Mariusz, et al. 2005. RHEED study of CdTe and HgCdTe thin films grown on Si by pulse laser deposition. *Thin Solid Films* 480:318–321.

Sathyamoorthy, R., Senthilarasu, Sundaram, Lalitha, S., Subbarayan, Arumugam, Natarajan, Krithikaa, and Mathew, Xavier. 2004. Electrical conduction properties of flash evaporated zinc phthalocyanine (ZnPc) thin films. *Solar Energy Materials and Solar Cells* 82 (1–2):169–177.

Shah, Arvind, Vallat-Sauvain, Evelyne, Torres, P., et al. 2000. Intrinsic microcrystalline silicon (μn-Si: H) deposited by VHF-GD (very high frequency-glow discharge): a new material for photovoltaics and opto-electronics. *Materials Science and Engineering: B* 69:219–226.

Sheikh, Arif D., Bera, Ashok, Haque, Md Azimul, et al. 2015. Atmospheric effects on the photovoltaic performance of hybrid perovskite solar cells. *Solar Energy Materials and Solar Cells* 137:6–14.

Solanki, Chetan Singh., Bilyalov, R.R., Poortmans, Jef, and Nijs, Jo. 2002. Transfer of a thin silicon film on to a ceramic substrate. *Thin Solid Films* 403:34–38.

Solar, Evergreen. 2021. Evergreen solar.

Sugathan, Vipinraj, John, Elsa, and Sudhakar, Kumarasamy. 2015. Recent improvements in dye sensitized solar cells: A review. *Renewable and Sustainable Energy Reviews* 52:54–64.

Sunpower. 2021. From Maxeon solar technologies. Available at: https://sunpower.maxeon.com/int/

Tang, Ching W. 1986. Two-layer organic photovoltaic cell. *Applied Physics Letters* 48 (2):183–185.

Tawada, Yoshihisa, and Yamagishi, Hideo. 2001. Mass-production of large size a-Si modules and future plan. *Solar Energy Materials and Solar Cells* 66 (1–4):95–105.

Vitanov, P., Delibasheva, M., Goranova, E., and Peneva, M. 2000. The influence of porous silicon coating on silicon solar cells with different emitter thicknesses. *Solar Energy Materials and Solar Cells* 61 (3):213–221.

von Roedern, Bolko, and Ullal, Harin S. 2008. The role of polycrystalline thin-film PV technologies in competitive PV module markets. *Paper read at 2008 33rd IEEE Photovoltaic Specialists Conference.* DOI: 10.1109/PVSC.2008.4922493

Wikipedia. 2020. Czochralski method.

Wurfel, Peter. 1982. The chemical potential of radiation. *Journal of Physics C: Solid State Physics* 15 (18):3967.

Xue, Xiaonan, Weng, Kangkang, Qi, Feng, et al. 2019. Steric engineering of alkylthiolation side chains to finely tune miscibility in nonfullerene polymer solar cells. *Advanced Energy Materials* 9 (4):1802686.

Yang, Jeffrey, Banerjee, Arindam, and Guha, Subhendu. 2003. Amorphous silicon based photovoltaics—hotovoltaicsicon based sr cellcib". *Solar Energy Materials and Solar Cells* 78 (1-4):597–612.

Zhu, Lei, Zhang, Ming, Zhou, Guanqing, et al. 2020. Efficient organic solar cell with 16.88% efficiency enabled by refined acceptor crystallization and morphology with improved charge transfer and transport properties. *Advanced Energy Materials* 10 (18):1904234.

Zhu, Lei, Zhong, Wenkai, Qiu, Chaoqun, et al. 2019. Aggregation-induced multilength scaled morphology enabling 11.76% efficiency in all-polymer solar cells using printing fabrication. *Advanced Materials* 31 (41):1902899.

8 Progress and Challenges of Semiconducting Materials for Solar Photocatalysis

Mridula Guin, Tanaya Kundu, Vinay K. Verma, and Nakshatra Bahadur Singh

CONTENTS

DOI: 10.1201/9781003188582-8

8.1 INTRODUCTION

The progress of human civilization is the motive behind energy crunch and environmental pollution. Development of industries in a fast rate and exploitation of fossil fuels are the two key reasons for environmental pollution. To combat the situation scientists are always in search of photocatalytic green technology. The observation of Fujishima and Honda (1972) about the photocatalytic decomposition of water using TiO_2 under sunlight establishes the era of semiconductor photocatalysis. Since then numerous investigations are being performed on photocatalysis for conversion of solar energy into sustainable and environment friendly energy. Photocatalysts are utilized for elimination of hazardous pollutants from water (Betancourt-Buitrago *et al.* 2019, Vaiano and Iervolino 2018), sterilization (Rong *et al.* 2018, Wang *et al.* 2017), hydrogen production (Chu *et al.* 2018, Zhang *et al.* 2019a, Wang *et al.* 2018), treatment of polluted air (Vidyasagar *et al.* 2019) and many other field of applications. Photocatalysis has immense potential to resolve the issues related to energy and environment. In spite of these potentials, the photocatalytic technology is far from industrial requirements. This may be due to non-feasibility of developing ideal photocatalysts with some special features that include large specific surface area, broad absorption range in sunlight, recyclability and high photocatalytic efficiency. When semiconductors are excited with light energy more than its band gap, energetic electron-hole pairs are created, which take part in the catalytic reaction. The electronic and optical characteristics of the semiconductors are upgraded for better performance in their response to visible light. In this chapter, optical properties of semiconductors, principle of semiconductor photocatalysis and types of semiconductor photocatalysts are discussed. In addition, methods for improving the efficiency of such catalyst are also discussed for various applications.

8.2 PRINCIPLES OF SEMICONDUCTOR PHOTOCATALYSIS

The physical properties of semiconductors are different than metals and dielectrics. The electronic band structures in different semiconductors display peculiarity. The conductivity of semiconductors falls between the metals and insulators. Semiconductors differ from the metals by their conductance behaviour with temperature. Resistance of semiconductors decreases exponentially with increasing temperature, which is just the opposite in the case of metals. Breaking of the bonds in the crystal creates free charge careers at high temperature. In addition to thermal motion, few other features such as source of radiation, electric field and existence of impurity affect the conductivity of the semiconductors.

Photocatalysis is the catalytic process in the presence of light, or in other words, we can say it is a light-induced catalytic process. Photocatalysis is a boon for mankind, which is used for solving environmental and energy-related issues. It is a low cost and highly competent method for decomposition and removal of pollutants, removal of pathogens, hydrogen production and air purification (Khan *et al.* 2015a). Fujishima and Honda (1972) have performed the reaction for decomposing water using titanium dioxide electrode in UV light, which established a new era of catalytic process. Since then a lot of extensive work has been performed for understanding the mechanism of photocatalysis and also to boost the competence of this process for diverse applications.

Photocatalysts can be sorted into two different classes: homogeneous and heterogeneous photocatalysts. Transition metal complexes are used as homogeneous photocatalysts while semiconducting materials are primarily used as heterogeneous photocatalysts. Semiconductors have some special characteristics that include beneficial electronic structures, light absorption characteristics, charge carrying properties and well-matched lifetime in the excited state. Heterogeneous photocatalysis is more preferred over the homogeneous photocatalysis as it is an economical method because of its requirement of ambient temperature and pressure conditions. A good semiconductor photocatalyst should maintain the following characteristics (Nakata and Fujishima 2012): (1) photo chemical activity, (2) absorb visible or UV light, (3) photostable and photon non-corrosive, (4) low cost, (5) safe for the environment and (6) chemically and biologically unreactive.

8.3 PHOTOCATALYTIC MECHANISM

In photocatalysis, semiconductors act as a catalyst by performing the role of sensitizers. The electronic structures of semiconductors are distinguished by their occupied valence band and unoccupied conduction band. The principle of photocatalytic process is based on stimulated redox reaction upon irradiation of light (Saravanan *et al.* 2017). The rate of the photocatalytic reaction reckons on the energy content of the photon and also the electronic nature of the semiconducting material. The mechanism of the process of photocatalysis by semiconductors is schematically presented in Figure 8.1. If the difference in energy of the valence band and the conduction band is equal or less than the absorbed energy of the photon of the irradiated light then the valence band electrons transit to the conduction band. As a result, holes are generated in the valence band. These freshly created holes have strong oxidizing ability. The reaction with water molecules can produce OH• radicals which are key performers in removal of pollutants. On the flip side, the electrons on the conduction band generate superoxide ions after reaction with dissolved oxygen. The holes and the electrons are the two partners in the redox reaction. The occurrence of consecutive oxidation/reduction reactions with the toxic substances, which are bound on the semiconductor surface, is the reason behind their degradation.

8.3.1 INFLUENCING FACTORS

Photocatalytic reactions are very complex in nature. It includes many intervening steps during the progress of the reaction. Several factors can alter the photocatalytic reaction mechanism and rate. Among them, the prime factors of immense importance are the nature of the catalyst, concentration of catalyst, surface properties of catalyst, reaction conditions, properties of light source, concentration of oxygen etc. playing a crucial role in influencing photocatalytic reaction. Various factors influencing photocatalytic processes are enlisted in Table 8.1.

8.3.1.1 Nature of a Catalyst

The rate of photocatalytic reaction is greatly influenced by the structure of the catalyst specially the shape, size and the surface area. The photocatalytic activity is influenced by the nature of the semiconductor. Among three distinct phases of TiO_2 (e.g. anatase, rutile and brookite), anatase is the most suitable for photocatalysis because of favourable band gap, stability, absorption power and

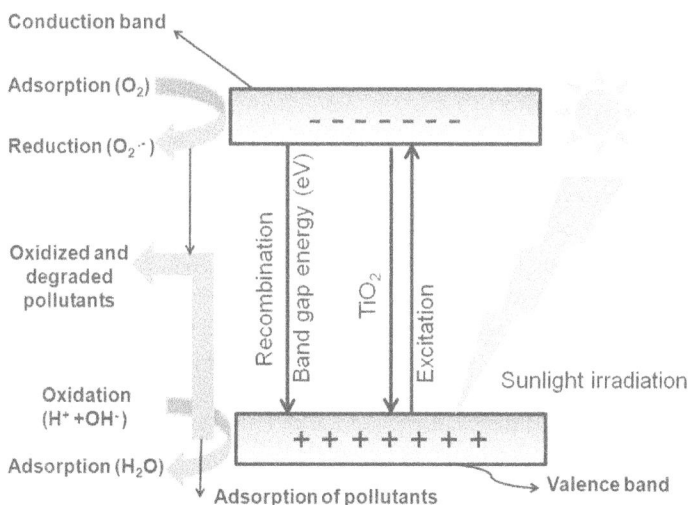

FIGURE 8.1 Photocatalytic mechanism in semiconductors.

TABLE 8.1

Influence of Various Factors in Photocatalytic Process

Influencing Factors		Effects on Photocatalysis
Concentration of	High	Rate of reaction increases
catalyst	Low	Rate of reaction decreases
Source of light	Source	Different wavelengths of light generated
	Intensity	Enhances photocatalytic processes
pH value		Variation of pH is related to degradation of the target
Inorganic ions	Anion	(a) Separation of photoelectrons and holes are maintained for improved catalytic efficiency
		(b) Acts as scavenger by reacting with hydroxyl radicals
	Cation	Photocatalytic degradation is affected by competitive adsorption over the active sites of catalysts
Temperature		Not much effect
Addition of oxidants		Recombination of photoelectrons and holes are restrained and better catalytic efficiency

faster hydroxylation (Gnanasekaran *et al.* 2015). Surface morphology of the semiconductor photocatalysts also play significant role in influencing the photocatalytic degradation of pollutants. It has been observed that spherical ZnO shows better catalytic efficiency than the rod shaped or spindle shaped ZnO because the former has large surface area (Saravanan *et al.* 2013). Furthermore, the nanosized TiO_2 exhibits better catalytic activity than the bulk TiO_2 because of increase in effective surface area. Due to this, the active sites on the surface becomes more and the rate of interfacial charge transfer improves and thereby catalytic activity gets enhanced (Han *et al.* 2014, Cernuto *et al.* 2011). The oxidation/reduction process of photocatalysis happens on the surface of the catalysts. Thus, the productivity of the catalyst is significantly dependent on the surface characteristics of the semiconductor catalysts (Khan *et al.* 2015b).

8.3.1.2 Temperature

A lot of research work has been performed to understand the effect of temperature on photocatalytic efficiency (Rehman *et al.* 2009). It has been observed that in TiO_2 material, the recombination process of electron-hole becomes faster at temperature above 80°C. In addition to that, desorption process becomes predominant to decrease the photocatalytic efficiency. Thus, the catalytic activity of TiO_2 is maximized between 20 and 80°C. The activation energy of photocatalytic degradation is very low (5–20 kJ/mol) and is effective at room temperature (Chatterjee and Dasgupta 2005).

8.3.1.3 pH

pH of the solution can alter the surface charge of the catalyst and thus influence the photocatalytic reaction remarkably. Many scientific reports have proven the effect of pH on the degradation of pollutants by photocatalytic mechanism (Kazeminezhad and Sadollahkhani 2016, Reza *et al.* 2015). It has been observed that organic dyes can be decomposed conveniently in the range of pH is 5–10. The efficiency reduced at the high acidic medium as well as high basic medium (Malato *et al.* 2009). The effect of pH is understood from the electrostatic interaction between pollutants and the charged particles at the surface. The extent of adsorption mostly relies on properties of surfaces that are modified by the change in pH of the solution.

8.3.1.4 Nature of Light

The photocatalytic reactions are dependent on various factors of light, e.g. source, wavelength, intensity etc. In general, the photocatalytic reaction rate escalates with the increase of intensity of the light. Furthermore, the final outcome of the photocatalytic reaction also alters with the light wavelength coming from different sources. TiO_2 absorbs the light in the UV region as it has a large band gap of 3.2 eV. The photocatalytic degradation of TiO_2 increases with increasing intensity up to 0–20 mW/cm^2. However, above the light intensity of about 25 mW/cm^2, it is observed that rate of reaction decreases due to the quicker recombination process between electrons and holes (Reza *et al.* 2015).

8.3.1.5 Concentration of a Catalyst

Increase in catalyst concentration leads to an enhanced rate of the reaction because of the presence of a large number of surface active sites. This facilitates the generation of more number of reactive hydroxyl free radical and superoxide free radicals, which in turn improves the photocatalytic degradation rate. In general, photocatalysis is directly dependent on concentration; however, it is realized that the rate of photocatalytic degradation reaction becomes unfavourable beyond the optimum concentration of the semiconductor catalyst (Malato *et al.* 2009). The reason behind this is that in very high concentration, the penetration of light inside the solution gets reduced.

8.3.1.6 Addition of Oxidants

Sometimes external oxidants or electron acceptors are added to reduce the recombination process among electron and hole so as to enhance photocatalysis efficiency. Radical formation has key role in decomposing the pollutants photolytically. Incorporation of a number of electron scavengers, e.g. H_2O_2, O_3, $KBrO_3$, $(NH_4)_2S_2O_8$ etc., have been investigated for the decomposition of various organic contaminants (Faisal *et al.* 2007). It has been observed that in most of the cases, the added oxidants enhanced the rate of catalysis and have a beneficial effect.

8.4 OPTICAL PROPERTIES OF SEMICONDUCTORS

Optical properties of semiconductors include a large number of phenomena that happen during the interaction of semiconductors with electromagnetic radiation (Singh 2007). The optical phenomena displayed by the semiconductors are dependent on the incident electromagnetic radiation, the reflected radiation and the light that passes through it (Nowak 2013). To understand various optical effects, few important characteristics are discussed as follows.

a. Reflection coefficient $R(\lambda)$

The reflection coefficient $R(\lambda)$ is the ratio of the reflected light to the incident light. It is given by the relation:

$$R(\lambda) = I(\lambda)_{ref} / I(\lambda)_{inc} \tag{8.1}$$

where $I_{ref}(\lambda)$ is the intensity of the reflected light and $I_{inc}(\lambda)$ is the intensity of the incident light. The reflection coefficient is dimensionless quantity and generally expressed in percentage. The relation of reflection coefficient with the component of complex refractive index is given by

$$R(\lambda) \approx \left[(n-1)^2 + k^2\right] / \left[(n-1)^2 + k^2\right]^2 \tag{8.2}$$

where n and k are real and imaginary part of the complex refractive index ($n^* = n - ik$). The parameters n and k are dimensionless and related with complex dielectric permittivity ($\varepsilon^* = \varepsilon' - \varepsilon''$).

Refractive coefficient, $n = (\varepsilon\mu^{1/2}) \approx \varepsilon^{1/2}$, as magnetic permeability $\mu \approx 1$ in the optical range.

Likewise absorption index $k = \varepsilon''/2n$

There are fundamental differences in light absorption in metals and semiconductors. Metals have quasi-continuous energy spectrum. Thus, photon energy is consumed to excite the free electrons within a few atomic layers. Accordingly, metals are characterized by light absorption on surface only. The absorption process is followed by the re-irradiation of the photons. Consequently metals exhibit high reflectivity of light (more than 90%).

On the other hand, the optical refractive index value of semiconductors comes around $n = 3$–4. In far infrared (IR) region, the semiconductor's reflection coefficient $R(\lambda)$ vary in the range of 25–40%. Interestingly in the visible range, fundamental absorption is observed where $k \gg n$. This is because the energy of the incident light is much higher than the band gap of the semiconductors. A huge number of excited free electrons are accumulated over the semiconductor's surface under visible light irradiation. The optical reflectance of semiconductors then increases and becomes close to 90%. Due to this, their opacity also increases and they show metallic luster. The light absorption in semiconductor is governed by the condition of $h\nu \geq E_g$ (E_g is semiconductor band gap). In general, the energy of the photon should be more than enough for ionizing the valence band electrons.

b. The absorption coefficient (α)

It is related to the absorption index k and wavelength of light λ by the equation $\alpha = 4\pi k / \lambda$. Furthermore, the absorption coefficient is characterized by Bouguer-Lambert's law, given by

$$I_x = I_{inc} \left(1 - R\right) e^{\alpha x} \tag{8.3}$$

or,

$$\alpha = x^{-1} \, ln\left\{\left[I_{inc} \left(1 - R\right)\right]/I_x\right\} \tag{8.4}$$

Bouguer-Lambert's law is related to light intensity absorbed by a sample with unit thickness. R is the reflection coefficient of the semiconductor. The parameter $(1-R)$ is the fraction of light that passed via the illuminated surface. When the light passed through a layered material, the flux intensity of the light reduced as it depends on the thickness. However, the loss of photons in the media does not depend on the light intensity and thickness of the absorbing layer. The dimension of absorption coefficient α is 1/length and has the unit cm^{-1}.

When light falls on the semiconductor crystal, the incident photons display different optical phenomena (Figure 8.2) by performing different physical and chemical transformation.

8.4.1　External Photoelectric Effect

External photoelectric effect is more critical in metals. In semiconductor, this effect is negligible under light irradiation in the optical range. In photoelectric effect, the electrons are knocked out from the valence band by the absorbed photon. A considerable amount of energy is desired as the

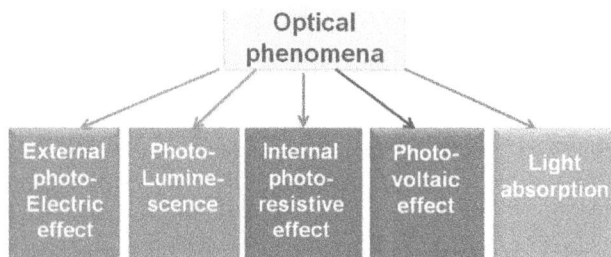

FIGURE 8.2　Various optical phenomena displayed by semiconductors.

energy of the ingested photons must be adequate to break the electrons from valance band and make them free. Accordingly, this phenomena in semiconductor is possible at light frequencies, considerably higher than visible light frequencies. But the incident solar radiation has very small proportion of high-frequency radiation leading to less chance of external photoelectric effect in conventional semiconductors.

8.4.2 PHOTOLUMINESCENCE

Photoluminescence is a secondary radiation phenomena mainly arising because of the system's nonequilibrium situation. Many devices are developed based on this phenomenon. When the semiconductor is irradiated with light, a pair of free electrons and holes is generated during the light absorption. These free electrons reach the excited state and remain there for a brief period of time known as lifetime. After completion of the lifetime, the charge carriers and the holes recombine and release energy in the form and radiation. This reradiation of photons while transiting from higher excited state to lower state is called as photoluminescence. Secondary radiation happens when the system comes back from the nonequilibrium state to the equilibrium state. Two types of luminescence, fluorescence and phosphorescence are observed. Fluorescence happens only during the excitement while phosphorescence continues for a bit more after excitation.

There are three different mechanisms by which photoluminescence occurs. The first mechanism is spontaneous mechanism in which the non-radiating transition from higher excited state to the lower excited state happens. After that further radiation leaves the molecule in the ground state. The attribute of this mechanism is the presence of the impurity state of the solid. The second mechanism is based on forced luminescence. This happens after the absorption of the light when the molecule reaches to some meta-stable state followed by transition to the other states. The third mechanism is the recombination luminescence in which the reassociation of the separated photoexcited particles occurs. This mechanism is also a good indicator of the presence of impurity centres in the semiconductor material. Therefore, the luminescence characteristics are important parameters of semiconductors that help in developing efficient photocatalysts.

8.4.3 INTERNAL PHOTOELECTRIC EFFECT

This effect is observed when photon energy is enough to break off the electrons from the valance band of the atoms and transfer them to conduction band. At that time, a large number of electron-hole pairs are illuminated and emerge close to semiconductor surface. In this case, the light is absorbed by intrinsic mechanism.

The investigation of the band structure in semiconductors is largely dependent on the absorption spectrum. The absorption of light is enhanced when the incident photon energies are either equal or more than the band gap of the semiconductors. It has been observed that during absorption, the impulse of photon is much smaller than the impulse of the electrons inside a crystal. Thus, the impulse of electron remains unaltered during absorption of photons.

Furthermore, as a consequence of light absorption, the generation of electron-hole combination increases in great number. Thus, the electrical conductivity of the semiconductor also increases accordingly to the large extent. This induced conductivity has an intrinsic quality as it is related to the energized condition of the semiconducting material. Light absorption in these solids is dependent on frequency of the light characterized by absorption coefficient (α).

8.4.3.1 Photoresistivity/Photoconductivity

Photoresistivity as well as photoconductivity is the result of the internal photoelectric effect. This phenomenon is dependent on the generation of light by the charge carriers due to intrinsic absorption or impurity absorption. After illuminating the semiconductor, light absorption takes place. In general, one electron-hole pair is created after the absorption of one photon. Because of the presence of a large number of photons in the light beam, electrical conductivity increases significantly during irradiation. Photoconductivity is an important quantity for solar photocatalysis and also for manufacturing electronic devices.

The total conductivity (σ_{ph}) because of light absorption after irradiation is presented as:

$$\sigma_{ph} = \sigma_0 + \Delta\sigma_{gen} = e\left(n_0 u_n + p_0 u_p\right) + e\ \Delta n\left(u_n + u_p\right)$$

In the above equation, σ_0 is known as dark conductivity or stationary conductivity which is inherent to the system. The second term $\Delta\sigma_{gen}$ is the conductivity that arises because of the light absorption. Δn is the density of charge carriers that are generated because of light absorption. Among all optical properties, photoconductivity is the most important for various applications. In the case of intrinsic semiconductors or weakly doped semiconductors, it is observed that $\Delta\sigma_{gen} \gg \sigma_0$.

The number of charge carriers that are created by light absorption is dependent on the value of quantum yield (η). For the intrinsic semiconductors, the quantum yield remains almost constant with value close to $\eta = 1$. This is the case which is verified up to certain low-frequency range at which photogeneration starts, up to certain low-frequency range. Thus, one electron-hole pair is generated by the absorption of one photon. The remaining energy is used for the creation of phonons. On the other hand, at higher energy of irradiated light more than one pair of charge carriers are generated by the absorption of one photon, and thus, $\eta > 1$. Excess energy in this case is utilized for valence bond interband-impact ionization and generation of extra electron-hole pairs. Temperature can also impact the quantum yield value. With increase of temperature the band gap decreases. Therefore, the interband ionization moves towards lower value of light energy.

Thus, the internal photoelectric effect of valence bond creates additional electron-hole pair which is known as secondary electron-hole pairs by ionization and yields $\eta > 1$. In this case, the photon energy is similar to violet light energy.

8.4.4 Photovoltaic Effect

It is generation of electromotive force by the dispersal of the charge by free electrons. As light cannot pass deep inside the semiconductor, it is absorbed by the surface layer. The average thickness of surface layer is equivalent to the path length of the photons. Therefore, a non-equilibrium dynamics between electron-hole pairs persist at the surface layer of semiconductor. According to Bouguer-Lambert law, the creation of this irregular charge carriers diminishes exponentially with time.

Due to the presence of excess amount of charge carriers, the dispersal of the electrons and holes are guided towards the depth of semiconductors. The wave current of electrons and holes proceed in the same direction. If the diffusion coefficients of electrons and holes are identical then there might be compensation between the streams. However, as the speed of electrons is more than the holes, the electroneutrality is not maintained. Consequently, diffusion coefficients of electrons are more, resulting in charge separation. The existence of space charge creates an electric field along the movement of the charges. A difference potential U_D is created between the bright surface and the dark (reverse) surface of the semiconductor because of the dynamic polarization of electrons and holes. If the thickness of the semiconductor is taken as d, the generated static electric field E_D (Dember field) is given by, $E_D = U_D/d$.

Thus, the photoelectric effect or Dember effect arises because of non-maintenance of electrical neutrality due to difference in the speed of electrons and holes.

8.4.5 Light Absorption Mechanism

The intensity of light decreases when it passes through the semiconductor because of the interaction with the matter. Consequently, it converts into the other forms of energy. The absorption coefficient usually is independent of intensity of light but depends on light wavelength.

Across the visible spectrum, the light wave interacts with the semiconductor crystal in different mechanism. Therefore, it exhibits features of spectral dependency throughout the spectrum. Light absorption is followed by transformation of photon energy into different other forms. The types of absorption mechanism are classified according to the energy states of the semiconductor lattice. Various types of absorption mechanism are (a) intrinsic, (b) interband, (c) impurity, (d) excitonic, (e) lattice and (f) plasma. A brief discussion on each mechanism is described in next section.

Intrinsic mechanism: The electronic transitions between allowed energy levels are known as fundamental or intrinsic absorption. In semiconductor it is the most important absorption mechanism near IR and visible region of the spectrum. This mechanism is governed by interaction between photons and electrons of own atoms of the crystal in the valance band. Photons with adequate energy can transfer energy to the electrons to remove them away from material and transport to higher energy levels. In intrinsic mechanism, photons are consumed by the semiconductor. The criterion for this transition to occur is $h\nu \geq E_g$ (where E_g is the band gap). The edge of the intrinsic absorption $\lambda_k = hc/E_g$ is calculated using the equality sign.

Interband absorption: When free charge carriers are present within the bands then interband absorption happens. During the interaction of photons with free in the allowed bands, the absorption occurs due to the transition of electrons or holes in allowed bands or sub-bands. The energy of the photon is utilized in moving the charge carriers in higher levels. Electric field of the light led to the synchronous oscillatory movements of the charge carriers. The excess energy accumulated by the charge carriers is released while colliding with lattice points.

Impurity absorption: This arises due to the movements of electrons or holes from the allowed band to the impurity level of the forbidden band. Because of the interaction of photon with the impurity atoms, ionization or excitation is the common observation in this case.

Excitonic absorption: A significant contribution to the absorption by the exciton generation is observed near the edge of fundamental absorption. The impurity atoms can impart a resonance character while interacting with photons. The randomly moving excitons i.e. the bound electron-hole pairs can absorb the photons. When the exciton collide with the impurity centre can either break upon into electron and holes by absorption of thermal energy or recombine and move to the excited state by radiating photon or transferring the energy as heat to the semiconductor lattice.

Lattice absorption or phonon absorption: It is also an important mechanism of light absorption in semiconductors. In the far-IR region, absorption by the crystal lattice which is known as phonon absorption or lattice absorption. When light energy absorbed by the vibrating atoms of the crystal, new phonons are generated. The phonon energy in the far-IR region overlaps partially with the absorption of free charge carriers.

Plasma absorption: It is the absorption of light energy by the electron-hole plasma for transition of plasma to higher energy quantum level. It happens in the presence of adequate high concentration of free charge carriers. It exhibits special feature of plasma resonance. The neighbourhood of plasma resonance has very high absorption coefficient. For this reason, it is usually explored a very thin reflection spectrum instead of absorption spectrum.

All absorption mechanisms are associated with absorption or emission of phonons. This is due to the conservation of impulse. Thus, light absorption mechanisms are amalgamation of different mechanism and are associated with electrons, holes and phonons.

8.5 TYPES OF SEMICONDUCTOR PHOTOCATALYST

8.5.1 OXIDE PHOTOCATALYST

Transition metal oxides of IVth period have unique physical and chemical properties. Existences of multivalence, high redox potential, high chemical resistance, stability at high temperature are few of those special physicochemical properties. They are abundantly available and less costly. Thus, they have high prospect for application in photocatalysis.

8.5.1.1 TiO$_2$-Based Photocatalysts

TiO$_2$ is the most popular photocatalyst for various applications such as degradation of organic pollutants, air purification, antibacterial disinfection etc. TiO$_2$ is available in three distinct crystalline forms: rutile, brookite and anatase. Among those rutile and anatase form is frequently used for photocatalytic purpose (Huang *et al.* 2015). The band gap of TiO$_2$ is 3.2 eV which enables it to absorb ultraviolet radiation. However, the ultraviolet radiation is only about 5% of the solar radiation. Thus, application of TiO$_2$ in visible range is limited. To make it an effective photocatalyst under visible radiation different modifications of TiO$_2$ has been adopted. The improvement of photocatalytic performance is carried out by cutting down the recombination of electrons and holes generated out of photon absorption. Various techniques such as surface modification by doping with metals/nonmetals, surface sensitization etc. are used commonly.

8.5.1.2 Bi$_2$O$_3$-Based Photocatalysts

Bismuth oxide is low-energy band gap (2.8 eV) material possessing high electrical conductivity. It has high potential in photocatalytic application because of its ability to absorb in the visible range. It has found wide application in photoelectric conversion, electrolyte material, superconducting material etc.

Bi$_2$O$_3$ exists in four different crystal structure forms that are α-Bi$_2$O$_3$, β-Bi$_2$O$_3$, γ-Bi$_2$O$_3$, δ-Bi$_2$O$_3$. The band gaps of α and β forms are 2.58 and 2.85 eV, respectively, leading to the absorption in the visible range with wavelength more than 400 nm. α-Bi$_2$O$_3$ is the most stable crystal structure among the four forms and has been investigated for its photocatalytic properties. Despite of having photocatalytic activity in the visible range, Bi$_2$O$_3$ has the limitations of fast recombination of charge carriers and holes. To improve catalytic efficiency, various modification methods are performed for further applications in water remediation, sterilization and purification process (Gao *et al.* 2015).

8.5.1.3 Other Oxide-Based Photocatalysts

In addition to TiO$_2$ and Bi$_2$O$_3$ there are many other metal oxides Fe$_2$O$_3$. ZnO and WO$_3$ which act as potential important photocatalysts. ZnO semiconductor has piezoelectric properties and can display quantum size effect. The band gap of 3.2 eV in ZnO is associated with absorption at 387 nm wavelength in UV range. The photocatalytic properties of ZnO are heavily reliant on surface architecture and the configuration of ZnO material. By altering the oxygen vacancies over the surface of the semiconductor material, the rate of the oxidation process can be adjusted. In turn the efficiency of the photocatalysis can be improved.

The excellent photocatalytic performance of WO$_3$ is because of its large specific surface area and superior adsorption power. Furthermore, it is a very stable metal oxide and is used for invisible material in various applications. The band gap of WO$_3$ is 2.8 eV which enables it to act as catalyst as well as co-catalyst in many processes.

Among iron oxide, Fe$_2$O$_3$ is the most investigated photocatalyst as it has strong adsorption in the visible region. It has a band gap of 2.2 eV allowing it to adsorb substantial amount of solar radiation. It is an n-type semiconductor. It has various photocatalytic applications, including decomposition of dyes (Ercan *et al.* 2015, Jablonski and Ranicke 2016) and reduction of metal ions such as silver (Sun *et al.* 2016).

8.5.2 Non-Oxide Photocatalysts

In addition to metal oxides, few non-oxides of transition metals also exhibit potential photocatalytic activity (Cheng *et al.* 2014, Chang *et al.* 2017, Khanchandani *et al.* 2012). In non-oxides sulphides and selenides of many metals received the special attention as photocatalysts. Their unique structure and physicochemical properties play key role in their catalytic activity. Among sulphide semiconductors, the most important are molybdenum sulphide (MoS$_2$), cadmium sulphide (CdS), tungsten sulphide (WS), copper sulphide (CuS) and zinc sulphide (ZnS). In addition to these, nitride-based catalyst, C_3N_4 is also an excellent photocatalyst. A brief discussion on the photocatalytic applications of these semiconductors are discussed in the following section.

8.5.2.1 CdS-Based Photocatalysts

CdS is a widely studied efficient photocatalyst in the range of visible light radiation (Zheng *et al.* 2019). The band gap of CdS is 2.42 eV exhibiting maximum absorption at 514 nm. Thus, the absorption wavelength range in the visible or ultraviolet region is below 514 nm. The conduction band energy (−0.90 eV) of CdS is in the lower edge in comparison to widely used photocatalyst TiO_2, ZnO etc. Thus, the photoelectrons generated during photocatalytic reaction of CdS, act as more powerful reducing agent. The band gap value of CdS is appropriate for many important photocatalytic reactions like decomposition of water, reduction of CO_2 etc. (Li *et al.* 2014). Despite these advantages, CdS has tendency for photo corrosion which limits the possibility of recovering the photocatalyst. Currently research is going on to reduce this fundamental problem to upgrade the efficient use of CdS photocatalyst. One of the most important measures to handle this issue is the synthesis of composite materials of CdS by ion doping method or by various other methods. It has been observed that the composite material has the capacity to utilize a wide spectrum of the visible light by absorbing longer wavelength electromagnetic radiation.

8.5.2.2 CuS-Based Photocatalysts

CuS is a p-type semiconductor. The band gap of 2.2 eV indicates that it can absorb visible light sufficiently. CuS semiconductor has exceptional optoelectronic properties, and thus, it has found applications in solar cells, photocatalysis, photothermal conversion, lithium battery electrodes and also in many other fields (Gorai *et al.* 2013). Because of large photocatalytic activity of CuS photocatalyst, it is utilized in the degradation of adsorbed organic matter from surface of a material. CuS materials are highly promising in photocatalysis process due to its suitable band gap and absorption wavelength (Mills *et al.* 2015).

8.5.2.3 ZnS-Based Photocatalysts

ZnS is a widely used high band gap ($E_g = 3.6$ eV) semiconductor. ZnS exists in two different forms of crystal structure, e.g. zinc blende (β-ZnS) and wurtzite (α-ZnS). At low temperature, zinc blende is the predominant form whereas at high temperature (more than 1024°C) α-form more stable. ZnS does not undergo oxidation and hydrolysis easily both in bulk and nanoscale. Thus, ZnS can act as an excellent photocatalyst. Easy fabrication method and environment benign nature made ZnS widely accepted in photocatalytic process. Various research groups are engaged in synthesizing nanoscale ZnS in the form of particles, nanowires, nanotubes and nanosheets (Khanlary *et al.* 2018, Reddy *et al.* 2019, Srinivasan *et al.* 2013, Wang *et al.* 2017).

The photocatalytic activity of ZnS is improved by adopting various methods such as by increasing surface area, doping with metals and non-metals and making composite or heterojunction. By altering the morphology of the ZnS, the surface to volume ratio can be enhanced (Srinivasan *et al.* 2013). The large surface area leads to more number of active sites on ZnS surface. Hence, the photocatalytic performance of ZnS increases upon increasing the specific surface area. When ZnS is doped with metals or non-metals its electronic properties and also the band structure changes. Because of this doping, absorption of ZnS under visible light is enhanced and catalytic activity is also improved. Additionally, controlling the electron-hole recombination process is also one of the important methods for improving the catalytic efficiency. Formation of heterojunction or making a composite structure reduces the recombination rate and thereby photocatalytic efficiency of ZnS is enhanced (Zhang *et al.* 2013).

8.5.3 NITRIDE-BASED PHOTOCATALYSTS

Carbon nitride, C_3N_4, is an attractive polymeric material having hardness more than diamonds (Cao *et al.* 2015). It can exist in different phases, e.g. α-, β-, cubic, quasi-cubic and graphitic phase. Graphitic carbon nitride, g-C_3N_4, have graphite like layered structure is a purely non-metallic

TABLE 8.2

Band Gaps of Some Selected Photocatalysts

Semiconductor	Conduction Band (CB) Energy in eV	Valence Band (VB) Energy in eV	Band Gap (E_g/eV)
CuO	−1.16	0.85	2.00
ZnO	−0.31	2.89	3.20
TiO_2 (anatase form)	−0.50	2.70	3.20
Fe_2O_3	0.28	2.48	2.20
WO_3	−0.10	2.70	2.80
Bi_2O_3	0.33	3.13	2.80
$BiVO_4$	−0.30	2.10	2.40
Ag_3PO_4 (cubic)	0.04	2.49	2.45
g-C_3N_4	−1.30	1.40	2.70
ZnS	−1.04	2.56	3.60
CdS	−0.90	1.50	2.40
Ta_3ON	−0.75	1.75	2.50
Ta_3N_5	−0.75	1.35	2.10

polymeric semiconductor (Fina *et al.* 2015). It has several unique properties, e.g. narrow band gap (2.7–2.8 eV), excellent thermal and chemical stability, fast electron-hole transporting ability. These features impart photocatalytic properties to g-C_3N_4. According to the band gap, g-C_3N_4 absorbs strongly around 400–450 nm in the visible range. Nowadays it is gaining much popularity as a photocatalyst because of its easily controllable structure. The properties can be tailored according to the role it plays. g-C_3N_4 has found extensive application as a photocatalyst in H_2 production, CO_2 reduction, photo hydrolysis, photodegradation of pollutants, disinfectants, production of capacitor material etc. (Cui *et al.* 2015, Liu *et al.* 2018, Lv *et al.* 2018, Maeda *et al.* 2014 Zhu *et al.* 2014,).

The band gaps of some selected photocatalysts are summarized in Table 8.2. The gap between conduction and valence band is the most important factor in photocatalytic applications.

8.6   CHALLENGES IN SEMICONDUCTOR PHOTOCATALYSIS

Semiconductor photocatalysts with moderate band gap of 1.1–3.8 eV are mostly investigated for solar photocatalysis (Figure 8.3) where various metal oxide, metal sulphide and metal selenide semiconductors are gaining much attention (Khan *et al.* 2015b). It has been observed that most sulphide and selenide semiconductors are unstable, toxic and photo corrosive. Thus, metal oxides such as TiO_2, ZnO, CeO_2 etc. are the most opted photocatalysts (Khan *et al.* 2014). Among all metal oxides TiO_2 is the most researched and well accepted photocatalyst for decomposing organic pollutants. It has band gap of 3.32 eV leading to the high photocatalytic efficiency (Schneider et *al.* 2014). Its high stability and non-toxic nature has made it to the most preferred photocatalyst. On the other hand ZnO and CeO_2 with similar band gap as of TiO_2, have high adsorption ability and thus preferred over TiO_2 (Choi *et al.* 2016). But photocatalytic activity of ZnO and TiO_2 is applicable at UV region as the large band gap makes them inactive under visible radiation. In the visible region, hematite can act as a potential photocatalytic material. But it has the disadvantage of formation of transient charge transfer states or corrosion property diminishing its catalytic efficiency. Sunlight consists of 46% visible, 47% IR and 5–7% ultraviolet (Khan *et al.* 2015b). Although TiO_2 and ZnO semiconductors inherently inactive for solar photocatalysis, a continuous effort is going on all over the world to develop technical methods for making them to absorb in the visible range of the solar radiation. In the next section few important methods for improving photocatalytic activity of these semiconductors are discussed.

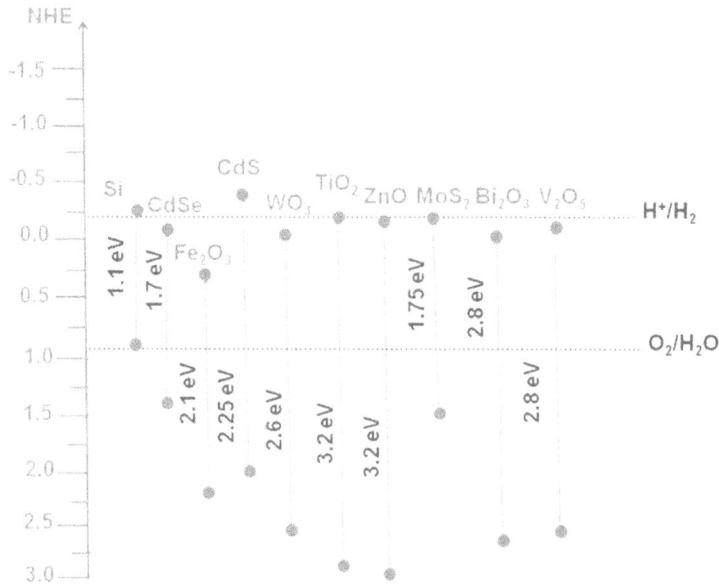

FIGURE 8.3 Band gap and redox potential of various semiconductors with respect to reference electrode normal hydrogen electrode (NHE).

8.6.1 Methods for Improving Photocatalytic Efficiency

Photocatalysis is environmentally safe method as it utilizes renewable solar energy and offers replacement for the conventional methods. Among its several advantages, the most important one is that it has wide range of applications without liberating toxic substances in the environment. Remediation of a large number of hazardous materials from waste water can be performed by photocatalytic method. Another important point is that it needs ambient environmental condition, small amount of chemical and less time for completion of the process. This method is highly efficient and the generation of secondary waste is minimal. In spite of these beneficial aspects of this method, there are certain limitations also. The semiconductor photocatalytic processes exhibits narrow absorption spectra which restrain their wide application possibility. In addition to that semiconductors exhibit low photon quantum efficiency and facile recombination process. The most effective way to improve the photocatalytic process is to diminish the rate of the recombination of photogenerated electrons and holes (Zhang *et al.* 2019b). Some important methods for improving the photocatalytic efficiency in semiconductor are discussed in the following section.

8.6.1.1 Deposition of Precious Metal

To prevent the recombination process, precious metal deposition is one of the best method. The principle of this method lies in suppressing the photogenerated electrons for further recombination by altering the semiconductor surface (Figure 8.4). The composite system of the semiconductor and the precious metal surface will accumulate excess positive and negative charges because of the disparity in values of the work function of the respective systems. Thus, the interface of the metal and the semiconductor exhibit band bending. Due to this, a Schottky barrier is generated which captures the excited electrons at the surface and reduce the recombination process. Overall an improved efficiency is observed in the system after precious metal deposition.

The most common precious metal used for this purpose is platinum (Liang *et al.* 2019). Some other noble metal, e.g. Ag, Au, Pd, Ru etc. are also reported for effective improvement of the photocatalytic performance. However, their utility is highly limited on the ground of high cost.

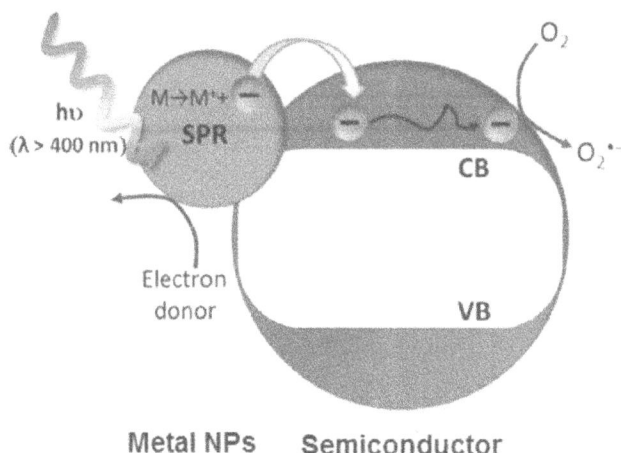

FIGURE 8.4   Scheme of metal deposition on semiconductor.

The principle of noble metal requires similar size of the metals. Nevertheless, due to size mismatch, direct penetration of noble metal is not much feasible into the semiconductor in doping process. Surface modifications of the semiconductor are performed by changing the electron distribution by depositing the noble metal over it. At the interface of the metal and semiconductor a space charge layer is formed due to the redistribution of the charge carriers. The recombination of photogenerated electrons and holes are suppressed because of the space charge layer favoured the easy movement of the electrons to the intersection area of noble metal/semiconductor. Thus, the efficiency of photocatalysis gets improved.

Various single metal (Ag, Pd and Pt) or bimetallic clusters (Ag/Pt, Ag/Pd and Pt/Pd) are used for the surface modifications on TiO_2 semiconductor for the decomposition of organic pollutants under UV-visible radiation (Klein *et al.* 2016). It has been observed that the degradation efficiency depends on the preparative method by which deposition has been performed. Gold nanoparticles deposited over sulphate TiO_2 surface have shown exceptional photocatalytic property by decomposing 97.6% Congo red dye (Zhang and Wang 2015). Padilla *et al.* (2016) deposit a number of noble metals in their single and complex form on TiO_2 by sonic degradation method. Among various photocatalysts they studied AuPd/TiO_2 is found to be the best for decomposing methyl orange dye. Silver metal being the less expensive and less toxic, is most favoured metal for depositing over the semiconductor surface for improved photocatalytic activity. Research reported by Jaafar *et al.* (2015) has shown that Ag-deposited TiO_2 photocatalyst can remove chlorophenol up to 94% at pH 5. This result is much higher than the photocatalytic ability of the other catalysts. The presence of certain amount of silver and oxygen vacancies suppresses the recombination of electron-hole pairs and enhances the photocatalytic efficiencies.

8.6.1.2   Semiconductor Composite

Blending of two semiconductors in such a manner so as to minimize the band gap is an important technique for improving photocatalytic efficiency. The composite photocatalyst can be prepared by doping, simple combination, out of phase combination or by multilayer structure. Presence of two different energy levels in the coupled system helps in separation of charges. The photogenerated electrons or holes of one semiconductor is moved to the conduction band/valence band of the partner semiconductor and thereby retards the process of recombination of charge carriers effectively.

A composite of cadmium sulphide with TiO_2 prepared by hot solvent method was found to act efficiently as photocatalyst for the removal of methyl orange, methylene blue and rhodamine B dyes (Arabzadeh and Salimi 2016). Wu *et al.* (2015) have synthesized $FeVO_4$-TiO_2 composite by blending nanorod $FeVO_4$ with TiO_2 by co-precipitation method. This composite can decompose NO gas 1.5 times faster than the nanoparticle $FeVO_4$-TiO_2 catalyst. The research work by Chu *et al.* (2016) have shown that composite of $MoSe_2$/TiO_2 is highly efficient towards reducing Cr^{6+} ions while the efficiency of $MoSe_2$ and TiO_2 separately as photocatalyst was much lesser. A ZrO_2-TiO_2 photocatalyst formulated by sol-gel method was investigated for the treatment of textile effluent (Das and Basu 2015). Shen *et al.* (2014a) have fabricated a porous composite by $InVO_4$ with nano-TiO_2 particles. Another work by the same group have reported g-C_3N_4-TiO_2 composite as highly effective photocatalyst (Shen *et al.* 2014b). These catalysts have the advantage of large size, porous nature, large surface area and narrow band gap. As more active sites are created over the surface, the lifetime of the electron-hole pairs increased and thus photocatalytic efficiency improves.

In addition to single component composite of TiO_2, there are reports of multi-component composite of TiO_2. Combination of multi-component with TiO_2 helps in improving light absorbing power and segregation of the pairs between electrons and holes. Composite of TiO_2 nanotubes with Pt and CdS nanoparticles have been reported by Gao *et al.* (2015). This photocatalyst effectively decompose methyl orange dye up to 91.9% under visible radiation. On the other hand the decomposition rate is much lesser with pure TiO_2 and single composite of Pt-TiO_2 and CdS-TiO_2 catalyst. The synergetic effect between CdS and Pt compliment the process of attraction of holes by CdS and transfer of photogenerated electrons by Pt. Thus, co-modified TiO_2 with CdS and Pt has higher photocatalytic efficiency due to the enhanced charge separation and fast photon capture process. A multilayer composite of TiO_2 nanotube prepared by Hui *et al.* (2017) has been found to decompose organic contaminants *p*-nitrophenol and rhodamine dye very efficiently.

8.6.1.3   Metal or Non-Metal Ion Doping

The performance of the semiconductor as photocatalyst can be improved by incorporating impurity in the forbidden band. This method is known to the researchers from long time and used as the most effective method. Although ion doping has the advantage of producing more semiconductor carriers; however, along with it there is the possibility of formation of ion traps to enhance the recombination of electron-hole pairs. The key role of ion doping is shifting of absorption wavelength of the semiconductor photocatalyst towards longer side so that the photoresponse can be stretched to the visible region. Ion doping is classified into three categories. They are (1) metal ion doping, (2) non-metal ion doping and (3) mixed doping.

8.6.1.3.1   *Doping with Metal Ion*

Doping with metal ion is the most frequently utilized technique for improving the photocatalytic performance. Semiconductors when doped with appropriate metal ion can display tremendous photocatalytic performance in the spectrum of visible light. The metal ion doping is performed by high-temperature firing or auxiliary deposition. It has been observed that doping with different transition metal ions, e.g. Fe, Co, Mn and Cu in ZnO lattice shrink the band gap in the visible region by spin exchange interactions (Kanade *et al.* 2007). Thus, the photocatalytic efficiency improved in the doped nanostructures. The Co doped ZnO nanostructure reported by Xiao *et al.* (2007) is highly efficient for degradation of methylene blue dye in visible region. Because of the incorporation of the transition metals into the semiconductor, the d-electron contour of the transition metals gets modified, and thus, the energy levels around the semiconductor lattice changes significantly (Ekambaram *et al.* 2007). Doping of vanadium ion with TiO_2 enables it to show photocatalytic activity at lower concentration because of absorption in the visible region. However, at high concentration, TiO_2 surface is clogged by the vanadium ion and suppresses photocatalytic activity (Yamashita *et al.* 2002). Chromium and vanadium ions when doped in TiO_2, the excitation wavelength expands till visible region and reaches near about 600 nm. Performance of the photocatalyst is found to vary with the nature of the doped transition metal ion and their concentration. Different metal ions exhibit different effects on photocatalytic performance. Doping with Fe^{3+} brings out improved photon efficiency. Wang *et al.* (2017) have synthesized Fe^{3+} doped TiO_2 nanotube by hydrothermal method.

This photocatalyst has shown remarkable photocatalytic performance in the field of dye removal and heavy metal reduction (Wang *et al.* 2017). Inturi *et al.* (2016) have investigated three different synthetic methodology for doping Cr^{3+} in TiO_2. Their observation indicates that Cr^{3+} doped TiO_2 prepared by flame spraying method is best for degrading chlorophenol in visible light. The reason of better photocatalytic performance compared with the other method is that by flame spraying method large surface area on the semiconductor surface are produced and thereby number of active sites are increased. Three different metal ions (Mg^{2+}, Zn^{2+} and Cr^{6+}) doped TiO_2 is prepared by co-precipitation method for the degradation of Congo red (Ma *et al.* 2017). MgZnCr-TiO_2 has shown excellent catalytic activity and recyclability up to five cycles.

In addition to transition metal ions, a number of rare earth metal ions are also doped in semiconductor material for improving photocatalytic performance. They have specific electronic structure different from transition metals. They have the ability to distort the semiconductor lattice, impart impurity defects and suppress the electron hole recombination. A number of rare earth ions such as La^{3+}, Ce^{3+}, Pr^{3+}, Nd^{3+}, Sm^{3+}, Eu^{3+}, Gd^{3+} etc. are doped in TiO_2 to understand their effect on photocatalysis (Villabona-Leal *et al.* 2015). Band gap and surface area of the doped nanostructures are found to be dependent on the nature and concentration of the lanthanoids. In contrary to transition metal ions, they do not influence the absorption in the visible range. Mixed doping with transition metals and rare earth metals of TiO_2 has been reported by Malengreaux *et al.* (2017). Among various samples of mixed metal oxides, La-Fe-TiO_2 was found to be the photocatalyst for the degradation of *p*-nitrophenol.

8.6.1.3.2 *Doping with Non-Metal Ion*

In spite of several advantages of the doping of transition metal ions in TiO_2, it is observed that few of them were thermodynamically unstable and enhances recombination process. When TiO_2 is doped with non-metals, a covalent bond is formed between the two. Doping helps in narrowing down the band gap and modification of the electronic and structural properties. Furthermore, oxygen defect sites are created as the non-metals replace the oxygen lattice sites in TiO_2 (Figure 8.5). The photocatalytic performance in visible region improved due to the synergetic effect (In *et al.* 2007). Different non-metals such as N, B, C, Si, S, P, Cl, F, I etc. are researched for doping with semiconductors for the potential improvement in photocatalytic performance. A lot of evidence can be obtained on doping of nitrogen in TiO_2 semiconductor for different photocatalytic applications (Castellanosleal *et al.* 2017, Horovitz *et al.* 2016, Lei *et al.* 2016). Studies by Shifu *et al.* (2009) have shown that nitrogen doped ZnO display red shift and better catalytic activity than the pure ZnO. Boron doped TiO_2 prepared

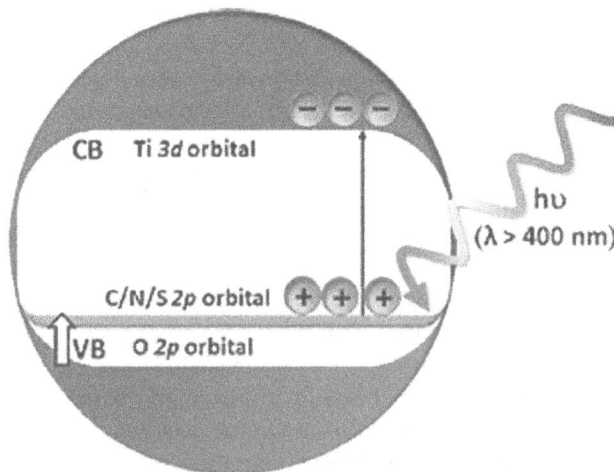

FIGURE 8.5 Displacement of valence band of TiO_2 by non-metal doping.

through hot solvent method was investigated by Simsek (2017) for the decomposition of dichloro-phenol and bisphenol A. Their observation indicates that under UV light, the catalytic activity of pure TiO_2 and B-TiO_2 are comparable for both the pollutants. However, under visible light irradiation B-TiO_2 acts far better catalyst for decomposition of both dichlorophenol and bisphenol A. This is mainly due to the doping with boron where forbidden band width of B-TiO_2 decreases and shifts the absorption towards visible range. Furthermore, the formation of B-O-Ti and B-O-B bonds enhances the stability and helps in trapping photogenerated electrons. Chitosan-TiO_2 composite prepared by hydrothermal method wherein C-doped TiO_2 acts as photocatalyst (He *et al.* 2017). The Ti-C chemical bond creates surface oxygen vacancies, which helps in adsorbing the organic contaminants. Doping with two non-metals, for example, N-S co-doped TiO_2 has been reported to have better photo degra-dation ability towards methyl orange in visible range because of higher oxygen vacancies (Wei *et al.* 2008). Another report by Ling *et al.* (2008) have demonstrated that the B and N co-doped TiO_2 is competent in absorbing visible light. This happens because of the change in the electronic structure of TiO_2 due to synergetic effect. C-N co-doped TiO_2 is synthesized via hydrothermal method (El-Sheikh *et al.* 2017). El-Sheikh *et al.* (2017) have shown that under visible light the C-N-TiO_2 display admirable photocatalytic performance by removing 100% ibuprofen while in the case of pure TiO_2 it was only 11.1%. Fluorine and graphene oxide co-doped TiO_2 is reported by Zhang *et al.* which has remarkable photocatalytic activity for bromate ion under UV irradiation (Zhang *et al.* 2017). Incorporation of gra-phene oxide imparts high conductivity as well as large specific surface area while fluoride ion doping improves the surface energy of 001 plane of TiO_2. These research works suggest that non-metal ion doping is also a powerful technique for boosting photocatalytic performance.

8.6.1.3.3 *Doping of Mixed Ion*

Mixed doping is the method in which doping is done by metal ions in combination with non-metal ions. This method has the advantage of both metals and non-metals simultaneously. Lei *et al.* (2017) have reported Fe-N-C co-doped TiO_2 for the removal of Cr^{6+} ions. The photocatalytic performance of Fe-N-C-TiO_2 was 100% while C-TiO_2 and N-C-TiO_2 was found to be much lower than this. A composite photocatalyst of TiO_2 is prepared by Chen and Liu (2017) by co-doping with Ce and N and diatomaceous earth particles. The 4f energy levels of Ce make the TiO_2 surface as trap for the photoelectron while nitrogen doping helps to reduce the forbidden band gap. The Ce-N-TiO_2 photo-catalyst is 100% effective in removing oxytetracycline within 240 min visible light irradiation. This result is much higher compared with the other catalyst used for the purpose. In the other report, Han *et al.* (2016) has prepared a three-element-doped mixed photocatalyst of Zr-Ni-N-TiO_2 for removal of NO and SO_2 gas under visible light. The result indicates that three element doping is more beneficial compared with the single or two element doped TiO_2 catalyst. The primary advantage of mixed dop-ing is to stretch the photocatalytic response towards the visible region to increase its performance.

8.6.1.4 Dye Photosensitization on Surface

Dye photosensitization method uses a photosensitizer to increase the photochemical response range of a semiconductor catalyst. Organic or inorganic chromophores are used as the photosensitizers. They strongly absorb in the visible region, and thus, the application of sensitized catalyst extends up to visible range from the UV radiation. The mechanism of dye sensitization is schematically shown in Figure 8.6. These dye sensitized semiconductor photocatalysts are extensively used for decom-posing organic contaminants. Various organic dyes, such as acid red 44, rhodamine B, eosin-Y, 8-hydroxyquinoline, merbromine and rhodamine 6G, have been used as sensitizers. Among inor-ganic dyes transition metal complexes of Ru and Pt is used to excite TiO_2 particles in visible light.

Porphyrin sensitized TiO_2/reduced graphene oxide (rGO) nanocomposite in rod shaped is prepared by Wan *et al.* (2016). The report suggests that this catalyst is more suitable for the decomposition of methylene blue dye under visible light. The degradation rate of this TiO_2/rGO nanorod composite is extremely high (92%) compared with the pure TiO_2 catalyst. In a separate work of Zoltan *et al.* (2016) prepared three asymmetric sensitizers 5-(*p*-nitrophenyl)-10,15,20-triphenylporphyrin with

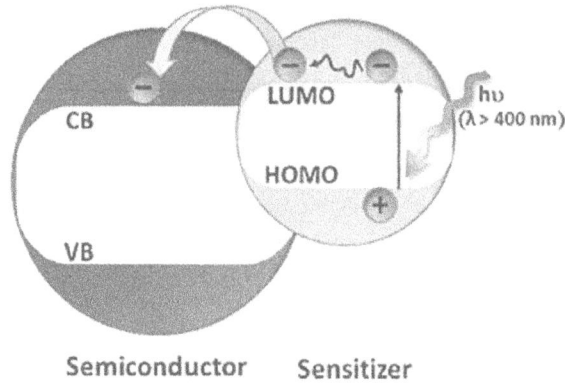

FIGURE 8.6 Mechanism of dye sensitization method.

Cu(II) and Zn(II) ions [Cu(II)-porphyrin and Zn(II)-porphyrin] have been prepared for sensitizing polycrystalline TiO_2 surface. The composite of TiO_2/Zn(II)-porphyrin displayed higher catalytic activity for degradation of Congo red dye than the other photocatalysts. TiO_2 surface is modified by attaching the carboxylate porphyrins of Co, Cu and Zn metal (Zhao *et al.* 2017). These sensitized photocatalysts have shown better catalytic activity for the decomposition of nitrophenol than pure TiO_2. Because of superb electron-accepting ability of Cu ions, it can separate electron-hole pairs better than the Co and Zn-porphyrin systems. Thus, Cu-porphyrin system was found to be the best for decomposing nitrophenol. Wei *et al.* (2016) have reported Cu(II)-tetrakis(4-carboxyphenyl) porphyrin with rGO sensitized TiO_2 catalyst for the degradation of methylene blue. The degradation rate of methylene blue was five times higher compared with pure TiO_2 and the catalytic activity can be recyclable up to six times. The band gap of the semiconductor can be altered by dye molecules and the photo response can be broadened. A report by Chowdhury *et al.* (2017) has shown reduction of band gap of TiO_2 by 1 eV by sensitizing with eosin dye. The dye injects the excited electrons into the conduction band of the semiconductor similar to the photogenerated electrons. Co-TiO_2 nano-doped materials are sensitized using metal-free complex phthalocyanine dye which is found to be excellent for the removal of pollutants (Altın *et al.* 2016). Because of the sensitization, the absorption wavelength shifted towards red side and the application of these catalysts extends up to visible region. Modification of the surface of TiO_2 nanoparticles by Cu(II)-phthalocyanine derivative reveal that the morphology of TiO_2 remains unaltered but the photosensitization improves the photo response range (Albay *et al.* 2016).

Table 8.3 lists some merit and demerit of various techniques applied for enhancement of photocatalytic efficiency.

TABLE 8.3
Merit and Demerit of Different Alteration Techniques for Semiconductor Photocatalyst

Method	Merit	Demerit
Doping of metal/non-metal	• Band gap decreases • Size of particle decreases	Leads to deformity
Depositing precious metal	Splitting of electron-hole becomes easier	Increases cost
Dye photosensitization on surface	Widen the range of light response	• Costly method • Chances of photolysis of dyes
Composite of semiconductors	• Electron-hole pairs complexes reduced • Widen the range of light response	Energy loss

8.7 CONCLUSIONS AND FUTURE PROSPECTS

Industrialization and high living standards of human being lead to the severe environment pollution along with energy crunch in all fields of society. In this respect, photocatalytic technology has evolved with promising features as environmentally friendly, economical and sustainable technology. In recent years, it has become a hot topic for research and development. Optical properties of semiconductors play important roles in photocatalysis. Different types of semiconductor such as oxide and non-oxide photocatalysts, sulphide and nitride-based photocatalysts and mechanism of photocatalysis have been discussed in this chapter. The efficiencies could be improved by blending and doping. Dye photosensitization method has also been used as a photosensitizer to increase the photochemical response range of a semiconductor catalyst.

Till date, the use of photocatalytic technology has not reach the point which is required for industrialization due to number of reasons. The photocatalysts developed so far suffer from various issues such as insufficient utilization of solar energy, marginal quantum efficiency, difficulty in recyclability and substandard photocatalyst efficiency. Therefore, there is a strong demand to develop powerful, productive, economical and stable visible light photocatalyst so that performance of photocatalytic activity enhances and photocatalytic technology gets boosted.

REFERENCES

Albay, C., Koç, M., Altın, İ., Bayrak, R., Değirmencioğlu, İ., Sökmen, M. 2016. New dye sensitized photocatalysts: Copper(II)-phthalocyanine/TiO$_2$ nanocomposite for water remediation. *J. Photochem. Photobiol., A*, 324, 117–125.

Altın, İ., Sökmen, M., Bıyıklıoğlu, Z. 2016. Sol gel synthesis of cobalt doped TiO$_2$ and its dye sensitization for efficient pollutant removal. *Mater. Sci. Semicond. Process.*, 45, 36–44.

Arabzadeh, A., Salimi, A. 2016. One dimensional CdS nanowire@TiO$_2$ nanoparticles core-shell as high performance photocatalyst for fast degradation of dye pollutants under visible and sunlight irradiation. *J. Colloid Interface Sci.*, 479, 43–54.

Betancourt-Buitrago, L.A., Ossa-Echeverry, O.E., Rodriguez-Vallejo, J.C., Barraza, J.M., Marriaga, N., Machuca-Martínez, F. 2019. Anoxic photocatalytic treatment of synthetic mining wastewater using TiO$_2$ and scavengers for complexed cyanide recovery. *Photochem. Photobiol. Sci.*, 18, 853–862.

Cao, S., Low, J., Yu, J., Jaroniec, M. 2015. Polymeric photocatalysts based on graphitic carbon nitride. *Adv. Mater.*, 46, 2150–2176.

Castellanosleal, E.L., Acevedopeña, P., Güizaargüello, V.R., Córdobatuta, E.M., Castellanosleal, E.L., Acevedopeña, P., Güizaargüello, V.R., Córdobatuta, E.M. 2017. N and F codoped TiO$_2$ thin films on stainless steel for photoelectrocatalytic removal of cyanide ions in aqueous solutions. *Mater. Res.*, 20, 487–495.

Cernuto, G., Masciocchi, N., Cervellino, A., Colonna, G.M., Guagliardi, A. 2011. Size and shape dependence of the photocatalytic activity of TiO$_2$ nanocrystals: A total scattering Debye function study. *J. Am. Chem. Soc.*, 133(9), 3114–3119.

Chang, L., He, X., Chen, L., Zhang, Y. 2017. Mercaptophenylboronic acid-capped Mn-doped ZnS quantum dots for highly selective and sensitive fluorescence detection of glycoproteins. *Sens. Actuators, B. Chem.*, 243, 72–77.

Chatterjee, D., Dasgupta, S. 2005. Visible light induced photocatalytic degradation of organic pollutants. *J. Photochem. Photobiol., C*, 6(2–3), 186–205.

Chen, Y., Liu, K. 2017. Fabrication of Ce/N co-doped TiO$_2$/diatomite granule catalyst and its improved visible light-driven photoactivity. *J. Hazard. Mater.*, 324, 139–150.

Cheng, J., Pan, Y., Zhu, J., Li, Z., Pan, J., Ma, Z. 2014. Hybrid network CuS monolith cathode materials synthesized via facile in situ melt-diffusion for Li-ion batteries. *J. Power. Sources*, 257, 192–197.

Choi, Y.I., Lee, S., Kim, S.K., Kim, Y., Cho, D.W., Khan, M.M., Sohn, Y. 2016. Fabrication of ZnO, ZnS, Ag-ZnS, and Au-ZnS microspheres for photocatalytic activities, CO oxidation and 2-hydroxyterephthalic acid synthesis. *J. Alloys Compd.*, 675, 46–56.

Chowdhury, P., Athapaththu, S., Elkamel, A., Ray, A.K. 2017. Visible-solar-light-driven photo-reduction and removal of cadmium ion with eosin Y-sensitized TiO$_2$ in aqueous solution of triethanolamine. *Sep. Purif. Technol.*, 174, 109–115.

Chu, H., Lei, W., Liu, X., Li, J., Zheng, W., Zhu, G., Li, C., Pan, L., Sun, C. 2016. Synergetic effect of TiO_2 as cocatalyst for enhanced visible light photocatalytic reduction of Cr(VI) on $MoSe_2$. *Appl. Catal., A Gen.*, 521, 19–25.

Chu, J., Han, X., Yu, Z., Du, Y., Song, B., Xu, P. 2018. Highly efficient visible-light-driven photocatalytic hydrogen production on $CdS/Cu_7S_4/g\text{-}C_3N_4$ ternary heterostructures. *ACS Appl. Mater. Interfaces*, 10, 20404–20411.

Cui, Y., Tang, Y., Wang, X. 2015. Template-free synthesis of graphitic carbon nitride hollow spheres for photocatalytic degradation of organic pollutants. *Mater. Lett.*, 161, 197–200.

Das, L., Basu, J.K. 2015. Photocatalytic treatment of textile effluent using titania–zirconia nano composite catalyst. *J. Ind. Eng. Chem.*, 24, 245–250.

Ekambaram, S., Iikubo, Y., Kudo, A. 2007. Combustion synthesis and photocatalytic properties of transition metal-incorporated ZnO. *J. Alloys Compd.*, 433(1–2), 237–240.

El-Sheikh, S.M., Khedr, T.M., Hakki, A., Ismail, A.A., Badawy, W.A., Bahnemann, D.W. 2017. Visible light activated carbon and nitrogen co-doped mesoporous TiO_2 as efficient photocatalyst for degradation of ibuprofen. *Sep. Purif. Technol.*, 173, 258–268.

Ercan, Ö., Deniz, S., Yetimoğlu, E.K., Aydın, A. 2015. Degradation of reactive dyes using advanced oxidation method. *Clean Soil Air Water*, 43, 1031–1036.

Faisal, M., Abu Tariq, M., Muneer, M. 2007. Photocatalysed degradation of two selected dyes in UV-irradiated aqueous suspensions of titania. *Dyes Pigments*, 72, 233–239.

Fina, F., Callear, S.K., Carins, G.M., Irvine, J.T.S. 2015. Structural investigation of graphitic carbon nitride via XRD and neutron diffraction. *Chem. Mater.*, 27, 2612–2618.

Fujishima, A., Honda, K. 1972. Electrochemical photolysis of water at a semiconductor electrode. *Nature*, 238, 37–38.

Gao, H., Wang, H., Jin, Y., Lv, J., Xu, G., Wang, D., Zhang, X., Chen, Z., Zheng, Z., Wu, Y. 2015. Controllable fabrication of immobilized ternary $CdS/Pt\text{-}TiO_2$ heteronanostructures toward high-performance visible light driven photocatalysis. *Phys. Chem. Chem. Phys.*, 17, 17755–17761.

Gao, M., Zhang, D., Pu, X., Ding, K., Li, H., Zhang, T., Ma, H. 2015. Combustion synthesis of Bi/BiOCl composites with enhanced electron–hole separation and excellent visible light photocatalytic properties. *Sep. Purif. Technol.*, 149, 288–294.

Gnanasekaran, L., Hemamalini, R., Ravichandran, K. 2015. Synthesis and characterization of TiO_2 quantum dots for photocatalytic application. *J. Saudi Chem. Soc.* 19(5), 589–594.

Gorai, S., Ganguli, D., Chaudhuri, S. 2013. Synthesis of copper sulfides of varying morphologies and stoichiometries controlled by chelating and nonchelating solvents in a solvothermal process. *Cryst. Growth Des.*, 5, 1–12.

Han, G., Wang, L., Pei, C., Shi, R., Liu, B., Zhao, H., Yang, H., Liu, S. 2014. Size-dependent optical properties and enhanced visible light photocatalytic activity of wurtzite CdSe hexagonal nanoflakes with dominant 001 facets. *J. Alloys Compd.*, 610, 62–68.

Han, Y., Zhang, J., Zhao, Y. 2016. Visible-light-induced photocatalytic oxidation of nitric oxide and sulfur dioxide: Discrete kinetics and mechanism. *Energy*, 103, 725–734.

He, D., Li, Y., Wang, I., Wu, J., Yang, Y., An, Q. 2017. Carbon wrapped and doped TiO_2 mesoporous nanostructure with efficient visible-light photocatalysis for NO removal. *Appl. Surf. Sci.*, 391, 318–325.

Horovitz, I., Avisar, D., Baker, M.A., Grilli, R., Lozzi, L., Di, C.D., Mamane, H. 2016. Carbamazepine degradation using a N-doped TiO_2 coated photocatalytic membrane reactor: Influence of physical parameters. *J. Hazard. Mater.*, 310, 98–107.

Huang, M., Yu, J., Li, B., Deng, C., Wang, L., Wu, W., Dong, L., Zhang, F., Fan, M. 2015. Intergrowth and coexistence effects of $TiO_2\text{-}SnO_2$ nanocomposite with excellent photocatalytic activity. *J. Alloys Compd.*, 629, 55–61.

Hui, F., Niu, T., Zhang, S., Bo, L., Cai, Q. 2017. Fabrication of layered $(CdS\text{-}Mn/MoS_2/CdTe)$-promoted TiO_2 nanotube arrays with superior photocatalytic properties. *J. Colloid Interface Sci.*, 486, 58–66.

In, S., Orlov, A., Berg, R., Garcia, F., Pedrosa-Jimenez, S., Tikhov, M.S., Wright, D.S., Lambert, R.M. 2007. Effective visible light-activated B-doped and B, N-codoped TiO_2 photocatalysts. *J. Am. Chem. Soc.*, 129(45), 13790–13791.

Inturi, S.N.R., Suidan, M., Smirniotis, P.G. 2016. Influence of synthesis method on leaching of the $Cr\text{-}TiO_2$ catalyst for visible light liquid phase photocatalysis and their stability. *Br. Med. J.*, 180, 351–361.

Jaafar, N.F., Jalil, A.A., Triwahyono, S., Efendi, J., Mukti, R.R., Jusoh, R., Jusoh, N.W.C., Karim, A.H., Salleh, N.F.M., Suendo, V. 2015. Direct in situ activation of Ag(0) nanoparticles in synthesis of Ag/TiO_2 and its photoactivity. *Appl. Surf. Sci.*, 338, 75–84.

Jablonski, M.R., Ranicke, H.B. 2016. Novel photo-fenton oxidation with sand and carbon filtration of high concentration reactive dyes both with and without biodegradation. *J. Text. Sci. Eng.*, 6, 2–17.

Kanade, K.G., Kale, B.B., Baeg, J.O., Lee, S.M., Lee, C.W., Moon, S.J., Chang, H. 2007. Self-assembled aligned Cu doped ZnO nanoparticles for photocatalytic hydrogen production under visible light irradiation. *Mater. Chem. Phys.*, 102, 98–104.

Kazeminezhad, I., Sadollahkhani, A. 2016. Influence of pH on the photocatalytic activity of ZnO nanoparticles. *J. Mater. Sci. Mater. Electron.*, 27(5), 4206–4215.

Khan, M.E., Khan, M.M., Cho, M.H. 2015a. Biogenic synthesis of a Ag–graphene nanocomposite with efficient photocatalytic degradation, electrical conductivity and photoelectrochemical performance. *New J. Chem.*, 39(10), 8121–8129.

Khan, M.M., Adil, S.F., Al-Mayouf, A. 2015b. Metal oxides as photocatalysts. *J. Saudi Chem. Soc.*, 19(5), 462–464.

Khan, M.M., Ansari, S.A., Pradhan, D., Ansari, M.O., Han, D.H., Lee, J., Cho, M.H. 2014. Band gap engineered TiO_2 nanoparticles for visible light induced photoelectrochemical and photocatalytic studies. *J. Mater. Chem. A*, 2(3), 637–644.

Khanchandani, S., Kundu, S., Patra, A., Ganguli, A.K. 2012. Shell thickness dependent photocatalytic properties of ZnO/CdS core–shell nanorods. *J. Phys. Chem. C*, 116, 23653–23662.

Khanlary, M.R., Alijarahi, S., Reyhani, A. 2018. Growth temperature dependence of VLS-grown ultra-long ZnS nanowires prepared by CVD method. *J. Theor. Appl. Phys.*, 12, 121–126.

Klein, M., Nadolna, J., Gołąbiewska, A., Mazierski, P., Klimczuk, T., Remita, H., Zaleska-Medynska, A. 2016. The effect of metal cluster deposition route on structure and photocatalytic activity of mono- and bimetallic nanoparticles supported on TiO_2 by radiolytic method. *Appl. Surf. Sci.*, 378, 37–48.

Lei, X.F., Zhang, Z.N., Wu, Z.X., Piao, Y.J., Chen, C., Li, X., Xue, X.X., Yang, H. 2017. Synthesis and characterization of Fe, N and C tri-doped polymorphic TiO_2 and the visible light photocatalytic reduction of Cr(VI). *Sep. Purif. Technol.*, 174, 66–74.

Lei, Z., Zhao, L., Li, M., Jin, Y., Song, W., Zeng, D., Xie, C. 2016. A modular calcination method to prepare modified N-doped TiO_2 nanoparticle with high photocatalytic activity. *Appl. Catal., B Environ.*, 183, 308–316.

Li, X., Wen, J., Low, J., Fang, Y., Yu, J. 2014. Design and fabrication of semiconductor photocatalyst for photocatalytic reduction of CO_2 to solar fuel. *Sci. China Mater.*, 57, 70–100.

Liang, Y., Ding, M., Yang, Y., Xu, K., Luo, X., Yu, T., Zhang, W., Liu, W., Yuan, C. 2019. Highly dispersed Pt nanoparticles on hierarchical titania nanoflowers with {010} facets for gas sensing and photocatalysis. *J. Mater. Sci.*, 54, 6826–6840.

Ling, Q., Sun, J., Zhou, Q. 2008. Preparation and characterization of visible-light-driven titania photocatalyst co-doped with boron and nitrogen. *Appl. Surf. Sci.*, 254(10), 3236–3241.

Liu, H., Zhang, Z.-G., He, H.-W., Wang, X.-X., Zhang, J., Zhang, Q.-Q., Tong, Y.-F., Liu, H.-L., Ramakrishna, S., Yan, S.-Y. 2018. One-step synthesis hetero structured g-C3N4/TiO2 composite for rapid degradation of pollutants in utilizing visible light. *Nanomaterials*, 8, 842–856.

Lv, Q., Si, W., He, J., Sun, L., Zhang, C., Wang, N., Yang, Z., Li, X., Wang, X., Deng, W. 2018. Selectively nitrogen-doped carbon materials as superior metal-free catalysts for oxygen reduction. *Nat. Commun.*, 9, 3376–3386.

Ma, C., Wang, F., Zhang, C., Yu, Z., Wei, J., Yang, Z., Li, Y., Li, Z., Zhu, M., Shen, L. 2017. Photocatalytic decomposition of Congo red under visible light irradiation using MgZnCr-TiO_2 layered double hydroxide. *Chemosphere*, 168, 80–90.

Maeda, K., Kuriki, R., Zhang, M., Wang, X., Ishitani, O. 2014. The effect of the pore-wall structure of carbon nitride on photocatalytic CO_2 reduction under visible light. *J. Mater. Chem. A*, 2, 15146–15151.

Malato, S., Fernández-Ibáñez, P., Maldonado, M.I., Blanco, J., Gernjak, W. 2009. Decontamination and disinfection of water by solar photocatalysis: Recent overview and trends. *Catal. Today*, 147(1), 1–59.

Malengreaux, C.M., Pirard, S.L., Léonard, G., Mahy, J.G., Herlitschke, M., Klobes, B., Hermann, R., Heinrichs, B., Bartlett, J.R. 2017. Study of the photocatalytic activity of Fe^{3+}, Cr^{3+}, La^{3+} and Eu^{3+} single-doped and co-doped TiO_2 catalysts produced by aqueous sol-gel processing. *J. Alloys Compd.*, 691, 726–738.

Nowak, M. 2013. Chapter 7: Optical properties of semiconductors. Roberto Murri (ed.), Silicon Based Thin Film Solar Cells, pp. 177–242. Bentham Science.

Mills, A., O'Rourke, C., Moore, K. 2015. Powder semiconductor photocatalysis in aqueous solution: An overview of kinetics-based reaction mechanisms. *J. Photochem. Photobiol., A*, 310, 66–105.

Nakata, K., Fujishima, A. 2012. TiO_2 photocatalysis: Design and applications. *J. Photochem. Photobiol., C*, 13(3), 169–189.

Padilla, R.H., Priecel, P., Lin, M., Lopezsanchez, J.A., Zhong, Z. 2016. A versatile sonication-assisted deposition-reduction method for preparing supported metal catalysts for catalytic applications. *Ultrason. Sonochem.*, 35, 631–639.

Reddy, P.L., Deshmukh, K., Chidambaram, K., Ali, M.M.N., Sadasivuni, K.K., Kumar, Y.R., Lakshmipathy, R., Pasha, S.K.K. 2019. Dielectric properties of polyvinyl alcohol (PVA) nanocomposites filled with green synthesized zinc sulphide (ZnS) nanoparticles. *J. Mater. Sci. Mater. Electron.*, 30, 1–12.

Rehman, S., Ullah, R., Butt, A.M., Gohar, N.D. 2009. Strategies of making TiO_2 and ZnO visible light active. *J. Hazard. Mater.*, 170(2–3), 560–569.

Reza, K.M., Kurny, A.S., Gulshan, F. 2015. Parameters affecting the photocatalytic degradation of dyes using TiO_2: A review. *Appl. Water Sci.*, 7, 1569–1578. doi:10.1007/s13201-015-0367-y

Rong, W., Zhang, W., Zhu, W., Yan, L., Li, S., Kai, C., Na, H., Suo, Y., Wang, J. 2018. Enhanced visible-light-driven photocatalytic sterilization of tungsten trioxide by surface-engineering oxygen vacancy and carbon matrix. *Chem. Eng. J.*, 348, 292–300.

Saravanan, R., Gracia, F., Stephen, A. 2017. Chapter 2: Basic principles, mechanism and challenges of photocatalysis. M.M. Khan *et al.* (eds.), Nanocomposites for Visible Light-induced Photocatalysis, Springer Series on Polymer and Composite Materials, pp. 19–40. Springer. doi:10.1007/978-3-319-62446-4_2

Saravanan, R., Gupta, V.K., Narayanan, V., Stephen, A. 2013. Comparative study on photocatalytic activity of ZnO prepared by different methods. *J. Mol. Liq.*, 181, 133–141.

Schneider, J., Matsuoka, M., Takeuchi, M., Zhang, J., Horiuchi, Y., Anpo, M., Bahnemann, D.W. 2014. Understanding TiO_2 photocatalysis: Mechanisms and materials. *Chem. Rev.*, 114(19), 9919–9986.

Shen, J., Yang, H., Feng, Y., Cai, Q., Shen, Q. 2014a. Synthesis of 3D hierarchical porous TiO_2/$InVO_4$ nanocomposites with enhanced visible-light photocatalytic properties. *Solid State Sci.*, 32, 8–12.

Shen, J., Yang, H., Shen, Q., Feng, Y., Cai, Q. 2014b. Template-free preparation and properties of mesoporous g-C_3N_4/TiO_2 nanocomposite photocatalyst. *CrystEngComm*, 16, 1868–1872.

Shifu, C., Wei, Z., Sujuan, Z., Wei, L. 2009. Preparation, characterization and photocatalytic activity of N-containing ZnO powder. *Chem. Eng. J.*, 148(2–3), 263–269.

Simsek, E.B. 2017. Solvothermal synthesized boron doped TiO_2 catalysts: Photocatalytic degradation of endocrine disrupting compounds and pharmaceuticals under visible light irradiation. *Appl. Catal., B Environ.*, 200, 309–322.

Singh, J. 2007. Electronic and Optical Properties of Semiconductor Structures. University of Michigan, Ann Arbor, Cambridge.

Srinivasan, N., Thirumaran, S., Ciattini, S. 2013. Preparation of ZnS nanosheets from (2,2′-bipyridine) bis (1,2,3,4-tetrahydroquinolinecarbodithioato-S,S′) zinc(II). *Spectrochim. Acta, A*, 102, 263–268.

Sun, Z., Li, L., Yang, H., Zhou, D., Li, H. 2016. Photocatalytic reduction of silver ion probe on nano titanium dioxide. *J. Xinyang Norm. Univ.*, 29, 79–83.

Vaiano, V., Iervolino, G. 2018. Facile method to immobilize ZnO particles on glass spheres for the photocatalytic treatment of tannery wastewater. *J. Colloid Interface Sci.* 518, 192–199.

Vidyasagar, D., Manwar, N., Gupta, A., Ghugal, S.G., Umare, S.S., Boukherroub, R. 2019. Phenyl-grafted carbon nitride semiconductor for photocatalytic CO_2-reduction and rapid degradation of organic dyes. *Catal. Sci. Technol.*, 9, 822–832.

Villabona-Leal, E.G., López-Neira, J.P., Pedraza-Avella, J.A., Pérez, E., Meza, O. 2015. Screening of factors influencing the photocatalytic activity of TiO_2: Ln (Ln = La, Ce, Pr, Nd, Sm, Eu and Gd) in the degradation of dyes. *Comput. Mater. Sci.*, 107, 48–53.

Wan, J., Wei, M., Hu, Z., Peng, Z., Wang, B., Feng, D., Shen, Y. 2016. Ternary composites of TiO_2 nanotubes with reduced graphene oxide (rGO) and meso-tetra (4-carboxyphenyl) porphyrin for enhanced visible light photocatalysis. *Int. J. Hydrogen Energy*, 41, 14692–14703.

Wang, Q., Jin, R., Zhang, M., Gao, S. 2017. Solvothermal preparation of Fe-doped TiO_2 nanotube arrays for enhancement in visible light induced photoelectrochemical performance. *J. Alloys Compd.*, 690, 139–144.

Wang, R., Shuang, N., Gang, L., Xu, X. 2018. Hollow $CaTiO_3$ cubes modified by La/Cr co-doping for efficient photocatalytic hydrogen production. *Appl. Catal., B Environ.*, 225, 139–147.

Wang, W., Li, G., Xia, D., An, T., Zhao, H., Wong, P.K. 2017. Photocatalytic nanomaterials for solar-driven bacterial inactivation: Recent progress and challenges. *Environ. Sci. Nano*, 4, 782–799.

Wang, Z., Zhang, H., Cao, H., Wang, L., Wan, Z., Hao, Y., Wang, X. 2017. Facile preparation of ZnS/CdS core/shell nanotubes and their enhanced photocatalytic performance. *Int. J. Hydrogen Energy*, 42, 17394–17402.

Wei, F., Ni, L., Cui, P. 2008. Preparation and characterization of N-S-codoped TiO_2 photocatalyst and its photocatalytic activity. *J. Hazard. Mater.*, 156(1–3), 135–140.

Wei, M., Wan, J., Hu, Z., Peng, Z., Wang, B. 2016. Enhanced photocatalytic degradation activity over TiO_2 nanotubes co-sensitized by reduced graphene oxide and copper(II) meso-tetra(4-carboxyphenyl) porphyrin. *Appl. Surf. Sci.*, 377, 149–158.

Wu, G., Li, J., Fang, Z., Lan, L., Wang, R., Gong, M., Chen, Y. 2015. $FeVO_4$ nanorods supported TiO_2 as a superior catalyst for NH_3–SCR reaction in a broad temperature range. *Catal. Commun.*, 64, 75–79.

Xiao, Q., Zhang, J., Xiao, C., Tan, X. 2007. Photocatalytic decolorization of methylene blue over $Zn_{1-x}Co_xO$ under visible light irradiation. *Mater. Sci. Eng., B*, 142(2–3), 121–125.

Yamashita, H., Harada, M., Misaka, J., Takeuchi, M., Ikeue, K., Anpo, M. 2002. Degradation of propanol diluted in water under visible light irradiation using metal ion-implanted titanium dioxide photocatalysts. *J. Photochem. Photobiol., A*, 148(1–3), 257–261.

Zhang, D., Wang, J. 2015. UV-visible light-activated Au@ pre-sulphated, monodisperse TiO_2 aggregates for treatment of Congo red and phthalylsulfathiazole. *J. Water Process Eng.*, 7, 187–195.

Zhang, F., Wang, X., Liu, H., Liu, C., Wan, Y., Long, Y., Cai, Z. 2019a. Recent advances and applications of semiconductor photocatalytic technology. *Appl. Sci.*, 9, 2489. doi:10.3390/app9122489

Zhang, H., Lin, J., Li, Z., Li, T., Jia, X., Wu, X.L., Hu, S., Lin, H., Chen, J., Zhu, J. 2019b. Organic dye doped graphitic carbon nitride with a tailored electronic structure for enhanced photocatalytic hydrogen production. *Catal. Sci. Technol.*, 9, 502–508.

Zhang, J., Wang, Y., Zhang, J., Lin, Z., Huang, F., Yu, J. 2013. Enhanced photocatalytic hydrogen production activities of Au-loaded ZnS flowers. *ACS Appl. Mater. Interfaces*, 5, 1031–1037.

Zhang, Y., Li, L., Liu, H., Lu, T. 2017. Graphene oxide and F co-doped TiO_2 with (001) facets for the photocatalytic reduction of bromate: Synthesis, characterization and reactivity. *Chem. Eng. J.*, 307, 860–867.

Zhao, X., Liu, X., Yu, M., Wang, C., Li, J. 2017. The highly efficient and stable Cu, Co, Zn-porphyrin-TiO_2 photocatalysts with heterojunction by using fashioned one-step method. *Dyes Pigments*, 136, 648–656.

Zheng, N.-C., Ouyang, T., Chen, Y., Wang, Z., Chen, D.-Y., Liu, Z.-Q. 2019. Ultrathin CdS shell-sensitized hollow S-doped CeO_2 spheres for efficient visible-light photocatalysis. *Catal. Sci. Technol.*, 9, 1357–1364.

Zhu, J., Xiao, P., Li, H., Carabineiro, S.A. 2014. Graphitic carbon nitride: Synthesis, properties, and applications in catalysis. *ACS Appl. Mater. Interfaces*, 6, 16449–16465.

Zoltan, T., Rosales, M.C., Yadarola, C. 2016. Reactive oxygen species quantification and their correlation with the photocatalytic activity of TiO_2 (anatase and rutile) sensitized with asymmetric porphyrins. *J. Environ. Chem. Eng.*, 4, 3967–3980.

9 Linear Optical Properties of Semiconductors
Principles and Applications

Muhammad Rizwan, Asma Ayub, Bakhtawer Razaq,
Aleena Shoukat, Iqra Ilyas, and Ambreen Usman

CONTENTS

DOI: 10.1201/9781003188582-9

9.1 INTRODUCTION

Semiconductors have become the top tier contributor to the revolution in the field of technology. The ever-growing research developments areas in semiconductors have made semiconductor materials covet for the technological. The revolution in research field has been made possible due to diverse properties of semiconductors (Stern 1963). A lot of work has been conducted toward understanding the properties of semiconductors and discovering new characteristics and responses to different factors, thus extending their applications (Stratton 2007). The major dive into semiconductors has led to the biggest discoveries in past, such as the discovery of transistor, tunneling and thus winning many Nobel Prizes. Understanding the basic properties of semiconductors has been the fundamental issue in solid-state physics and basic material science (Peter and Cardona 2010). A comprehensive knowledge of the optical properties is necessary to lead to a better understanding of behavior of semiconductor materials.

The basic understanding of the band structure is fundamental in optical properties of semiconductors. In traditional physics, most physical systems have been studied by manipulating their optical properties or using some optical probes such as photons. Understanding of the optical properties thoroughly necessitated to make the physicists aware of the underlying phenomenon in solid materials (Schmitt-Rink, Chemla, and Miller 1989).

Optical properties of semiconductors are very closely connected to their electronic band structure. The band structure is determined by the crystallography of the atoms of solids. Optical properties are just the interaction between electromagnetic radiation and the semiconductor and this interaction leads to properties like absorption, reflection, refraction, polarization and scattering (Alonso and Garriga 2018). The complete explanation of optical properties requires a proper modeling based on the interaction taken into consideration. Optical properties of semiconductors can be explained using a semiclassical model or considering the microscopic details of the interaction using a quantum mechanical model (Klingshirn 2012b).

Linear optical properties are those properties that appeared when the electromagnetic and semiconductor interactions happen at low-lying energy levels. That is why these properties depend on the wavelength of the light. Linear optical properties also depend on the material purity, external factors such as temperature and pressure. Furthermore, introducing impurities in the form of doping also affects the properties.

Linear optical properties can be characterized as intrinsic and extrinsic linear optical properties. Intrinsic linear optical properties are related to lattice defects and interband transitions, while extrinsic linear properties of semiconductors are related to crystal defects. In this chapter, we will review linear optical properties in detail step-by-step starting with establishing background to optical properties. In the first half of the chapter, we will establish the mathematical explanation using classical electrodynamics, and in the next half, we will explain the linear optical effects one by one.

9.2 ELEMENTARY THEORY OF SOLIDS

In this section, we will discuss the properties of electromagnetic waves, since their behavior is crucial to understanding optical properties.

9.2.1 Maxwell's Equations

The theoretical framework of optical properties of solids can be well explained by Maxwell's equations such as (Maxwell 1865):

$$\nabla . \mathbf{B} = 0 \tag{9.1}$$

$$\nabla . \mathbf{D} = \rho \tag{9.2}$$

TABLE 9.1

Quantities Used in Maxwell's Equations and Their Respective Units

Quantity	MKS Units	Practical Units
Electric Field (E)	1 V/m	10^{-2} V/cm
Electric displacement (D)	1 C/m^2	10^{-4} C/cm^2
Charge density (ρ)	1 C/m^3	10^{-6} C/cm^3
Current density (J)	1 A/m^2	10^{-4} A/cm^2
Magnetic induction (B)	1 Wb/m^2	10^4 G
Magnetic field (H)	1 A/m	$4\pi \times 10^{-3}$ Oe

$$\mu_0 = 4\pi \times 10^{-7}\,\text{H/m} = 1.25664 \times 10^{-6}\,\text{H/m}$$

$$\epsilon_0 = \left(\mu_0 C^2 \right)^{-1} = 8.8542 \times 10^{-12}\,\text{F/m}$$

$$\left(\frac{\mu_0}{\epsilon_0} \right)^{(1/2)} = \mu_0 c = \left(\epsilon_0 c \right)^{-1} = 376.73\,\text{ohm}$$

$$\nabla \times \mathbf{E} = -\frac{\partial \mathbf{B}}{\partial t} \tag{9.3}$$

$$\nabla \times \mathbf{H} = \mathbf{j} + \frac{\partial \mathbf{D}}{\partial t} \tag{9.4}$$

where **B** is magnetic field, **D** is electric displacement, **E** is electric field and **H** is magnetization, whereas **j** is electric current density (Band 2006). Quantities in Maxwell's equations and their units are provided in Table 9.1 (Stern 1963).

The beauty of Maxwell's equations lies in the way that these incorporate the effect of the medium into their character. Thus, we can fully comprehend the behavior of light in matter and vacuum. Maxwell's equations are modified in the presentence of medium as follows (Equations 9.5–9.8) (Stöcker 1999)

Equations (9.1–9.4) are modified according to requirements of environment:

$$\nabla . \mathbf{B} = 0 \tag{9.5}$$

$$\varepsilon_0 \nabla . \mathbf{E} = \rho - \nabla . \mathbf{P} \tag{9.6}$$

$$\nabla \times \mathbf{E} = -\frac{\partial \mathbf{B}}{\partial t} \tag{9.7}$$

$$\nabla \times \mathbf{B} - \frac{c^2 \partial \mathbf{E}}{\partial t} = \mu_0 \left(\mathbf{J}_{\text{ext}} + \mathbf{J}_{\text{cond}} + \frac{\partial \mathbf{P}}{\partial t} + \nabla \times \mathbf{M} \right) \tag{9.8}$$

where **P** is polarization vector. Relationship between **P** and **D** is given by

$$\mathbf{D} = \varepsilon_0 \mathbf{E} + \mathbf{P} \tag{9.9}$$

$$\mathbf{H} = (\mu_0^{-1} \mathbf{B}) - \mathbf{M} \tag{9.10}$$

In Equations (9.9) and (9.10), **P** is polarization and **M** is magnetization. In Equation (9.6), the charge density has an extra term; similarly in Equation (9.8), current density j has an external current density $\mathbf{J}_{\text{ext}}$, conduction current density $\mathbf{J}_{\text{cond}}$ and two extra terms, but the other two

Equations (9.5) and (9.7) are free from source terms indicating that the electromagnetic fields arise from changes and currents and no magnetic charge, since there is no magnetic charge present in nature. So magnetic field **B** is source free (Band 2006).

In order to complete the formalism of classical electromagnetism for optical properties of solids, we will also include the Lorentz force as represented by Equation (9.11),

$$\mathbf{F} = q(\mathbf{E} + \mathbf{v} \times \mathbf{B}) \tag{9.11}$$

Equation (9.2) shows that electric displacement **D** comes from electric charge density ρ, whereas magnetic induction is source free as shown by Equation (9.5) and Equation (9.6) indicates how varying electric and magnetic fields generate each other. This is one of the significant outcomes of Maxwell's equations. From Equations (9.9) and (9.10), we can see that electric displacement is the combination of electric field and polarization but the magnetic field density is composed of magnetic field and magnetization. The main difference between **B** and **H** is that **B** is source free, while **H** is not.

Microscopic Maxwell's equations consider all charges such as bound electrons and protons as source of the electric field given by $\mathbf{E}_{\mathbf{micro}}$, where the charges represent the ρ_{bound}. For $\mathbf{H}_{micro}$, both current density **j** and all spins serve as source. Microscopic Maxwell's equations can be transformed into macroscopic equations by averaging the quantities over small volumes and thus changing ρ_{bound} by $-\nabla \cdot \mathbf{P}$ and **j** bound by $\mathbf{P} + \text{curl } \mathbf{M}(\mathbf{r}, t)/\mu_0$ (Meschede 2007).

Maxwell's equations are modified in matter and vacuum because behavior of electromagnetic radiation in matter and vacuum is different. Next, we will examine the behavior of Maxwell's equations in vacuum and matter to obtain the wave equations in vacuum and matter.

9.2.1.1   Behavior of Electromagnetic Radiations in Vacuum

In vacuum, the following conditions are satisfied as given by

$$P = 0;$$
$$M = 0;$$
$$\rho = 0;$$
$$j = 0; \tag{9.12}$$

where **P** is polarization, **M** is magnetic induction, **j** is electric current density and ρ is charge density.

Applying these conditions to Maxwell's equations as given by Equations (9.1–9.4), we get the following wave equation in vacuum:

$$\nabla^2 \mathbf{E} - \mu_0 \varepsilon_0 \ddot{\mathbf{E}} = 0 \tag{9.13}$$

Solution of the second-order differential wave equation in wave Equation (9.13) is given by Equation (9.14) and is applicable to all electric fields:

$$\mathbf{E}(\mathbf{r}, t) = \mathbf{E}_0 f(\mathbf{k}.\mathbf{r} - \omega t) \tag{9.14}$$

where E_0 is the amplitude, and **f** is an arbitrary function for which the second derivative exists. ω is angular frequency and **k** is wave vector and relation between the two quantities is given by Equation (9.15) as follows:

$$\frac{\omega}{\mathbf{k}} = \left(\frac{1}{\mu_0 \varepsilon_0}\right)^2 = c, k = 2\pi/\lambda \tag{9.15}$$

For a plane harmonic wave, Equation (9.14) becomes

$$\mathbf{E}(\mathbf{r},t) = \mathbf{E}_0 \exp[i(\mathbf{k}.\mathbf{r} - \omega t)] \qquad (9.16)$$

Similarly, we can get an expression for H using the vacuum conditions as given in Equation (9.12). The substitution of plane wave solution in Maxwell's equations gives Equation (9.17):

$$\mathbf{H} = (\omega\mu_0)^{-1}\mathbf{k} \times \mathbf{E} = \mathbf{H}_0 \exp[i(\mathbf{k}.\mathbf{r} - \omega t)] \qquad (9.17)$$

where $\mathbf{k}$ is wave vector and H_0 is amplitude of magnetic field and is given by Equation (9.18) as

$$H_0 = (\omega\mu_0)^{-1}k \times E_0 \qquad (9.18)$$

By using Equations (9.9), (9.10) and (9.12), we can write the wave equations in vacuum as Equations (9.16) and (9.17):

$$\mathbf{D} = \mathbf{D}_0 \exp[i(\mathbf{k}.\mathbf{r} - \omega t)] = \varepsilon_0 \mathbf{E}_0 \exp[i(\mathbf{k}.\mathbf{r} - \omega t)] \qquad (9.19)$$

$$\mathbf{B} = \mathbf{B}_0 \exp[i(\mathbf{k}.\mathbf{r} - \omega t)] = \omega^{-1}\mathbf{k} \times \mathbf{E}_0 \exp[i(\mathbf{k}.\mathbf{r} - \omega t)] \qquad (9.20)$$

From Equations (9.19) and (9.20), the following conclusions can be drawn:

$$\mathbf{D} \perp \mathbf{k}, \mathbf{k} \perp \mathbf{B}, \mathbf{B} \perp \mathbf{D}. \qquad (9.21)$$

If medium is isotropic in vacuum, the following relationships can be obtained:

$$\mathbf{E}\|\mathbf{D} \text{ and } \mathbf{H}\|\mathbf{B}. \qquad (9.22)$$

The momentum density (momentum per unit volume) $\mathbf{\Pi}$ of the electromagnetic field is given by

$$\mathbf{\Pi} = \mathbf{D} \times \mathbf{B} \qquad (9.23)$$

$\mathbf{\Pi}\|\mathbf{k}, \mathbf{\Pi}$ is parallel to $\mathbf{k}$. And the energy flux density of the electromagnetic radiation represented by the Poynting vector $\mathbf{S}$ is

$$\mathbf{S} = \mathbf{E} \times \mathbf{H} \qquad (9.24)$$

$\mathbf{S}$ is parallel to $\mathbf{\Pi}$ in vacuum and isotropic materials. Poynting vector $\mathbf{S}$ is the function of constantly varying space and time (Haug and Koch 2009; Saleh and Teich 2019).

9.2.1.2 Behavior of Electromagnetic Radiations in Matter

Now we will exploit the behavior of light in matter and see how Maxwell's relations hold their shape and character in matter. For semiconductors, we make the following assumption that there is no external charge density ($\rho = 0$).

Then the wave equation for electric field in matter is modified as follows:

$$\nabla^2\mathbf{E} - \mu_0\varepsilon_0\ddot{\mathbf{E}} = \mu_0\ddot{\mathbf{P}} + \mu_0\dot{\mathbf{J}} + \nabla \times \dot{\mathbf{M}} \qquad (9.25)$$

Comparing Equations (9.13) and (9.25), we can clearly see how presence of medium modifies Maxwell's equations according to behavior of light. From this comparison, we also confer that Equation (9.25) is a homogenous analogous equation of Equation (9.13) that indicates that the sources of electromagnetic radiation can be varying current density, curl of varying **M** or varying dipole moment or polarization **P** for which the second derivative exist. The second assumption that we make is that our material is nonmagnetic, so the third term vanishes in Equation (9.25). All semiconductors have a certain level of diamagnetic behavior. But even in paramagnetic and ferromagnetic materials, the character disappears for higher frequencies, so Equation (9.25) becomes

$$\nabla^2 \mathbf{E} - \mu_0 \varepsilon_0 \ddot{\mathbf{E}} = \mu_0 \ddot{\mathbf{P}} + \mu_0 \dot{\mathbf{J}} \tag{9.26}$$

The presence of impurities affects the optical properties of semiconductors (Dietl 1994; Furdyna 1988; Klingshirn 2012b). Thus, optical properties are affected differently for pure or weakly or strongly doped semiconductors. In case the semiconductor is pure or has light or weak doping, carrier density **j** is small, consequently so is the conductivity, so we can make the following assumption:

$$|\mathbf{j}| = |\sigma \mathbf{E}| \ll |\mathbf{D}| \tag{9.27}$$

where | | in Equation (9.27) indicates magnitude and **j** is as follows:

$$\mathbf{j} = \sigma \mathbf{E} \tag{9.28}$$

And σ is conductivity.

But this relationship (Equation (9.27)) is not true for heavily doped semiconductors, since carrier density **j** is high, so is conductivity, σ, which will affect the optical properties. Based on all previous assumptions, we can finally write the linear wave equation in matter:

$$\nabla^2 \mathbf{E} - \mu_0 \varepsilon_0 \ddot{\mathbf{E}} = \mu_0 \ddot{\mathbf{P}} \tag{9.29}$$

From the previous equation, we get the fact that every dipole moment **p**, and every polarization **P**, will radiate an electromagnetic radiation, given that their second derivative in time exists (Stöcker 1999).

Since we are discussing linear optical properties, we will consider linear optics here and make some assumptions to establish a linear relationship between electric displacement (**D**), electric field **E** and polarization **P** as given next:

$$\frac{\mathbf{P}}{\varepsilon_0} = \chi \mathbf{E} \tag{9.30}$$

or

$$\mathbf{D} = \varepsilon_0 (1 + \chi) \mathbf{E} = \varepsilon \varepsilon_0 \mathbf{E} \tag{9.31}$$

with $\varepsilon = \chi + 1$, where ε and χ are the linear response functions known as dielectric function and linear susceptibility. Every linear response function has imaginary and real part that gives information about the linear optical properties of semiconductors. Both quantities are functions of ω, the angular frequency. Relationships between linear response functions and their dependence on angular frequency are given in Equations (9.32–9.34).

$$\varepsilon = \varepsilon(\omega, \mathbf{k}), \varepsilon \text{ is function of } \omega \text{ and k} \tag{9.32}$$

$$\chi = \chi(\omega, \mathbf{k}) = \varepsilon(\omega, \mathbf{k}) - 1, \tag{9.33}$$

$$\varepsilon(\omega, \mathbf{k}) = \varepsilon_1(\omega, \mathbf{k}) + i\varepsilon_2(\omega, \mathbf{k}) \tag{9.34}$$

Value of $\varepsilon(\omega)$ for $\omega = 0$ is called dielectric constant that is an important optical constant of semiconductors. Linear response functions are tensor of rank two but we consider them as scalar quantities (Stößel 2013).

Based on the previous linear optics relations, we can modify the wave equation for a dielectric medium, given as next:

$$\nabla^2 \mathbf{E} - \mu_0 \varepsilon_0 \varepsilon(\omega) \ddot{\mathbf{E}} = 0 \tag{9.35}$$

where $\varepsilon(\omega)$ is considered to be spatially constant. This wave equation in dielectric medium can be further modified for magnetic materials as follows:

$$\nabla^2 \mathbf{E} - \mu(\omega) \mu_0 \varepsilon_0 \varepsilon(\omega) \ddot{\mathbf{E}} = 0 \tag{9.36}$$

where $\mu(\omega)$ is the magnetic permeability.

Finally, the solution of linear wave equation in matter is given by

$$\mathbf{E} = \mathbf{E}_0 f(\mathbf{k}.\mathbf{r} - \omega t), \tag{9.37}$$

For plane harmonic wave,

$$\mathbf{E} = \mathbf{E}_0 \exp[i(\mathbf{k}.\mathbf{r} - \omega t)]. \tag{9.38}$$

These are the same solutions that obtained for vacuum as given in Equations (9.14) and (9.16), but the relationship between angular frequency ω and wave vector $\mathbf{k}$ can now be written as

$$\frac{c^2}{\omega^2} k^2 = \varepsilon(\omega) \tag{9.39}$$

Contrary, the relationship between angular frequency ω and wave vector $\mathbf{k}$ in vacuum is given by Equation (9.15), where k can be written as

$$\mathbf{k} = \frac{\omega}{c} \sqrt{\varepsilon(\omega)} = \frac{2\pi}{\lambda_{vac}} \sqrt{\varepsilon(\omega)} = \mathbf{k}_{vac} \sqrt{\varepsilon(\omega)} \tag{9.40}$$

We can write $\sqrt{\varepsilon(\omega)}$ as function of $n(\omega)$, which is the complex refractive index of the materials and can be expanded as

$$n(\omega) = n(\omega) + i\kappa(\omega) = \sqrt{\varepsilon(\omega)} \tag{9.41}$$

The final form of linear wave equation for electric field in matter (Equation (9.38)) can now be written using Equations (9.40) and (9.41) as

$$E = E_0 \exp\left[i(\mathbf{k}.\mathbf{r} - \omega t)\right] = E = E_0 \exp\left[i\left(\frac{\omega}{c}n(\omega)k'.r - \omega t\right)\right] \exp\left[i\left(-\frac{\omega}{c}n(\omega)(k'.r - \omega t)\right)\right] \tag{9.42}$$

where $n(\omega)$ and k' explain the propagation of light in matter and the second argument in exponential explains attenuation of light and often accounts for absorption and is called extinction coefficient $\kappa(\omega)$. Comparing Equation (9.42) with the law of absorption, we get the absorption coefficient as follows:

$$\alpha(\omega) = \frac{2\omega}{c} \mathbf{H}(\omega) = (4\pi/\lambda_{vac}) \mathbf{H}(\omega) \tag{9.43}$$

Similarly, solution of linear wave equation for magnetic field part in matter can be written as

$$\mathbf{H}_0 = (\omega\mu_0)^{-1}\mathbf{k}\times\mathbf{E}_0. \tag{9.44}$$

The difference in wave equations obtained in matter and vacuum from Maxwell's equations is that we get longitudinal solutions along with transverse wave solutions in case of medium, which we did not obtain in vacuum.

For longitudinal waves obtained in case of matter, we have

$$\mathbf{E}\|\mathbf{k}$$

which gives $\mathbf{H} = \mathbf{0}$, and magnetic field $\mathbf{B} = \mu_0\mathbf{H} = 0$. The longitudinal waves obtained in the case of electromagnetic radiation interaction in matter are not electromagnetic radiation but are polarization waves for which E and P are opposite and the fields such as $\mathbf{B}$, $\mathbf{H}$ and $\mathbf{D}$ are vanishing terms (Kalt 2004; Pendry 2000; Smith, Pendry, and Wiltshire 2004).

9.3   LINEAR OPTICAL EFFECTS IN SEMICONDUCTORS

When light interacts with matter, the electromagnetic wave or photon interacts with atoms or molecules of a semiconductor material; this interaction can result in two possible cases. One is the case in which the photons of incident light are absorbed by the material leading to an excitation of atoms to higher energy levels in semiconductor. For such kind of interaction, the bandgap of the semiconductor will be less or more than the incident photon energy. The photon that the atom absorbed can decay in many ways. One way is that the atoms can react via the spontaneous emission. In spontaneous emission, the atoms emit the absorbed photons, but photons have different energies with respect to incident photon (Schneider 2004).

The second case is when the incident light does not have enough energy, as in the photon, energy is less than bandgap, then it will not cause any excitation within material; it will simply lead to disturbances in the material. The disturbance of charges leads to acceleration of charges. As we know, any accelerated charged particles radiate electromagnetic waves, so these charges will emit secondary electromagnetic radiation (Van Vechten and Keller 1980).

The radiation emitted will have the same energy as the incident photon but will have different phase; this is the basic principle behind every linear optical effect such as reflection, absorption and scattering (Fox 2001).

To understand the interaction between an optical wave and medium of semiconductor, we need to review the linear wave equation. As in the previous section, we discussed Maxwell's equations and their formulation in vacuum and medium. Maxwell's equations enabled us to establish a relationship between electrical and magnetic phenomenon. In this section, we will go back to Maxwell's equations and explain mathematical formulation with respect to linear optics.

Reproducing Maxwell's equations (as given in Equations (9.1–9.4)), where $\mathbf{E}$ and $\mathbf{H}$ are electric and magnetic field vectors, respectively, and $\mathbf{D}$ and $\mathbf{B}$ are corresponding electric and magnetic field densities, respectively. The electrical density is represented by j. In case of vacuum, the relationship between these quantities is given by

$$\mathbf{D} = \varepsilon_0\mathbf{E} \tag{9.45}$$

and

$$\mathbf{B} = \mu_0\mathbf{H} \tag{9.46}$$

The modified Maxwell's equations in vacuum are given by

$$\nabla \times \mathbf{E} = -\frac{\partial \mathbf{B}}{\partial t} \tag{9.47}$$

$$\nabla \times \mathbf{B} = \mu_0 \varepsilon_0 \frac{\partial \mathbf{E}}{\partial t} \tag{9.48}$$

We know that $\mathbf{B}$ is source free as per the previous discussion; after substituting the value of $\mathbf{B}$ in Equation (9.47), the following wave equation resulted:

$$\nabla \mathbf{E} = \frac{1}{c^2}\frac{\partial^2 \mathbf{E}}{\partial t^2} \tag{9.49}$$

Now in case of medium, the relationship between $\mathbf{D}$ and $\mathbf{E}$ is modified by a material-dependent property called relative permittivity, ε_r

$$\mathbf{D} = \varepsilon_r \varepsilon_0 \mathbf{E} \tag{9.50}$$

Relative permittivity is related to dielectric susceptibility χ, and is always propotional to polarization in linear optics, so $\mathbf{D}$ becomes

$$\mathbf{D} = (1 + \chi)\varepsilon_r \varepsilon_0 \mathbf{E} = \varepsilon_0 \mathbf{E}\chi = \varepsilon_0 \mathbf{E} + \mathbf{P} \tag{9.51}$$

Polarization $\mathbf{P}$ is always a linear function of electric field vector $\mathbf{E}$.
Wave equation in medium is now written as

$$\nabla \mathbf{E} = \frac{1}{c^2}\frac{\partial^2 \mathbf{E}}{\partial t^2} + \varepsilon_0 \frac{1}{c^2}\frac{\partial^2 \mathbf{P}}{\partial t^2} \tag{9.52}$$

Equation (9.52) is the modified form of Equation (9.49). From Equation (9.52), we confer that an electromagnetic wave in medium will have two parts, a primary (first term) and secondary part (second term), and both will superimpose one another. The secondary part that is the second term arises from the polarization of the medium due to the presence of an external electric field. The secondary wave will have a phase shift relative to primary waves; thus, waves travel slower in the presence of medium. Solution of linear wave equation when the material is insulator is given by

$$\mathbf{E}(\mathbf{z},t) = (1/2)\hat{\mathbf{E}}\,\exp\left[j\left(\hat{\mathbf{n}}\mathbf{k}\mathbf{Oz} - \omega t\right) + \text{c.c}\right]\exp(i) = |\hat{\mathbf{E}}|\,\mathrm{Cos}\left(\hat{\mathbf{n}}\mathbf{k}\mathbf{Oz} - \omega t + \varphi_0\right)\exp[i] \tag{9.53}$$

$$\hat{\mathbf{E}} = \mathrm{Re}\left(\hat{\mathbf{E}}\right) + j\,\mathrm{Im}\left(\hat{\mathbf{E}}\right) = \mathbf{a} + \mathbf{bj} \tag{9.54}$$

The solution given in Equation (9.54) consists of real and imaginary parts that account for different linear optical effects in solids. This solution does not hold for nonlinear optics, as superposition rule does not apply there due to the complex nature of solution (Schneider 2004).

9.4   ELECTRONIC BAND STRUCTURE

The electronic band structures of semiconductors basically arise from the interchange between atoms. It is critical to see that there are two sorts of commitments to these collaborations. In the first place, the nature of the atoms decides their interaction and, second, the spatial order presents a regulation that advances or confines that strength contingent upon the symmetry of the ordering (Schwoerer and Wolf 2007). The two commitments that figure out as a result of the steadiest design are reflected in the electronic band structure, that is, the conceivable electronic energy states as an element of the wave vector (Köhler and Bässler 2015).

Each solid has its own distinctive energy-band structure. This variety in band structure is answerable for the wide scope of electrical and optical properties distinguishable in different materials. In semiconductors, the electrons are restricted to various groups of energy and prohibited from different areas (Trügler 2016). The energy contrast between the maxima of valence band (VB) and the minima of conduction band (CB) is referred to as electronic bandgap. Electrons can hop starting with one band then onto the next. Though all together for an electron to hop from a VB to a CB, it requires a particular least amount of energy for this change (Kittel, McEuen, and McEuen 1996). The necessary required energy varies with various materials. Electrons can acquire sufficient energy to leap to the CB by retaining either a photon (light) or phonon (heat) (Probert 2011).

According to literature, the semiconductor materials have non-zero or intermediate bandgap. At 0K temperatures, semiconductors show the behavior as an insulator. At moderate temperature, which is beneath the melting point, they permit the electron's thermal excitation into the CB (Dean 1984).

9.4.1 Direct and Indirect Bandgap

Bandgaps are typically divided into two groups on the basis of momentum distribution. If the momentum of the material is the same in the maxima of VB and in the minima of CB, then it is referred to as the direct bandgap of the material. Otherwise, the material has indirect bandgap if both do not have the same momentum (Rizwan et al. 2021). For those materials that have direct band, a photon that has larger energy than the bandgap is required for the direct transition of electron from the VB to CB. But in other case, those materials that have indirect bandgap, both a photon and a phonon are required for the transition of electron from valence to CB (Gu et al. 2007). Both types of materials have some applications. The materials having direct bandgap are used in light emitting diodes (LEDs), photovoltaics (PVs) and laser diodes and the other materials having indirect bandgap are commonly utilized in PVs and LEDs if these materials have some other advantageous characteristics (Yuan et al. 2018).

9.4.2 Assumptions and Limits of Band Structure Theory

There are some presumptions that are fundamental for band hypothesis to be effective:

- **Infinite size system:** For the continuous band, the part of material should comprise an enormous number of particles. Since a naturally visible piece of material consists of 10^{23} molecules, this is certifiably not a genuine limitation; band hypothesis even relates to atomic estimated semiconductors in integrated circuits. With adjustments, the idea of band hypothesis can likewise be stretched out to frameworks that are just "enormous" along certain dimensions, like two-dimensional electron system (Zhang et al. 2012).
- **In homogeneities and interfaces:** Close to surfaces, intersections and difference in homogeneities, the bulk band arrangement is disturbed. Not exclusively are there any nearby limited scope disturbances, yet in addition to neighborhood charge inequality characteristics. These charge irregular characters have electrostatic impacts that broaden intensely into semiconductors (Voon and Willatzen 2009).
- **Homogeneous System:** Band structure is a characteristic property of a material, which accepts that the material is homogeneous. For all intents and purposes, this implies that the substance faces of the material should be uniform all through the part.
- **Small Systems:** There is no possibility of continuous band structure for those frameworks that are small along each measurement (like quantum dot) (Glazer, Burns, and Glazer 2012).

9.4.3 ENERGY CONSIDERATIONS

When singular atoms interact to form the solid, then energy bands are developed through the atomic energy levels. The bonds between molecules are principally framed by the external valence electrons due to the fact that all the inward electrons are more firmly joined to the nucleus. Thus, a solid may be considered to be made up due to the core electrons and the outermost electrons. We considered a carbon atom as an example to represent the most photovoltaic semiconductor. Its electronic configuration in VB is $2s^2$, $2p^2$ (Aspnes and Yoo 1999).

The covalent bonding in the tetrahedral ordering of carbon atom can be made up in diamond structure due to having four valence electrons through the sp^3 hybridization of the outermost orbitals in the bonding procedure. That is the origin of the best inorganic semiconductors, among which are Ge, Si and furthermore compound semiconductors of the wurtzite, chalcopyrite structures and zinc blende (Cohen and Chelikowsky 2012). To form π bonds, the non-hybridized p orbitals from the neighboring carbon atoms are overlapped and also taking into account the delocalization of electrons, which is conjugation, and these types of bonds are necessary for organic semiconductors (Fox 2001).

The energy difference depends upon two parameters; one is bond length and the other is iconicity of the bonds. The bond length is inversely related to the bandgap. If the bond length of one atom is larger than the other, then its energy will be smaller. But for some compound semiconductors, it also depends upon the iconicity of the bonds. For those compounds having the same bond length, for example, Ge and GaAs, the second one has larger gap because of the iconicity of bond's contribution in GaAs (Faber and Malloy 1992).

9.4.4 DIELECTRIC FUNCTION

There is a close relationship between dielectric function and band structure of a semiconductor. Dielectric function can be examined experimentally or by using suitable models. By using experimental method, data are analyzed and then measurements are carried out and the dielectric function is calculated (Shindo, Morita, and Kamimura 1965).

Similarly, the dielectric function can be parameterized by using appropriate models according to the physical properties of the studied materials.

The interaction between the electrons and the electromagnetic field of a plane wave can be calculated as $\mathbf{p} + e\mathbf{A}$, here, the vector potential $\mathbf{A}$, of the electromagnetic field within the light wave vector q and frequency ω, is given as $\mathbf{A} = \hat{e} A_0 \exp(i\mathbf{qr} - \omega t)$, and the polarization vector is $\hat{e}$ (中山隆史 1989).

$$H_{1e}\Psi_n(r) = \left[\frac{p^2}{2m} + V(r)\right]\Psi_n(r) = E_n\Psi_n(r) \tag{9.55}$$

The resultant Hamiltonian consists of the two time-dependent additional terms and the unperturbed term. The interaction between the electrons and the radiation is described by the two time-dependent additional terms. To acquire the direct optical response, neglect the slighter quadratic term $e^2 A^2/(2m)$ and keep the prevailing term $(e\mathbf{A} \cdot \mathbf{p})/m$.

From the VB to CB, the transition probability per unit time, W, for an electronic transition is given by using the time-dependent perturbation theory as

$$W(v, c, \mathbf{k}) = \frac{e^2 A_0^2}{m^2}\left|M_{cv}(\mathbf{k}, \mathbf{k}')\right|^2 \frac{2\pi}{\hbar}\delta\left(E_{ck'} - E_{vk} - \hbar\omega\right) \tag{9.56}$$

where, $M_{cv}(\mathbf{k}, \mathbf{k}')$ is matrix element, given as

$$M_{cv}(\mathbf{k}, \mathbf{k}') = \left\langle c\mathbf{k}'\left|\hat{e}.\mathbf{p}e^{i\mathbf{qr}}\right|v\mathbf{k}\right\rangle$$

Equation (9.56) is called Fermi's Golden rule.

By relating the selection rules of space group, the calculation of matrix element $M_{cv}(\mathbf{k}, \mathbf{k}')$ is simplified. For illustration, wave vector conservation $\mathbf{k}' = \mathbf{k} + \mathbf{q} + \mathbf{g}$ is required by translational symmetry. It is due to that, in the optical range, the reciprocal lattice vector is $\approx 10^{-4}$ times greater than the light wave vector; we have $\mathbf{k}' = \mathbf{k}$ in Equation (9.56) and this situation is responsible for the direct interband alteration (Peter and Cardona 2010). In certain semiconductors, indirect interband transition takes place, the basic absorption edge is assumed by methods in which phonons take part. Here, $\mathbf{k}' = \mathbf{k} + \mathbf{q} \pm \mathbf{Q} + \mathbf{g}$, i.e., $\mathbf{k}' = \mathbf{k} \pm \mathbf{Q}$, is used for indirect interband transition, where the wave vector of the emitted and absorbed photon is represented by Q.

By multiplying the transition rate with $\hbar\omega$ and summing previous all changes that occur between the filled |v> states and unfilled |c> states for completely $\mathbf{k}$ vectors of the Brillouin zone, the absorbed energy per unit time and per unit volume can be obtained (Ibach and Lueth 2009). The imaginary part of the dielectric function can be computed by the absorption coefficient $\alpha(\omega)$ and it is attained by the consequence of dividing this sum with the incident energy:

$$\varepsilon_2(\omega) = \frac{\pi e^2}{3\varepsilon_0 m^2 \omega^2} \sum_{CV} \int \frac{2\mathrm{d}k}{(2\pi)^3} |M_{cv}(k)|^2 \, \delta(E_{ck} - E_{vk} - \hbar\omega) \tag{9.57}$$

The previous equation can be assessed from the electronic structure hypothesis techniques, and additionally, the total dielectric function can be determined by the Kramers-Kronig relationship. In Equation (9.57), the volume integral over the BZ converted into a surface integral above the surface of constant energy variance, $\hbar\omega = E_{ck} - E_{vk}$, by the δ function (Fox 2001). The matrix component M_{cv} could be used as constant by seeing the involvement in ε_2 of an insignificant time period of BZ. At that point, $\varepsilon_2(\omega)$ is corresponding to the joint density of states (JDOS):

$$\varepsilon_2(\omega) \propto J_{cv}(\omega) = \frac{2}{(2\pi)^3} \int \frac{\mathrm{d}S}{\left| \nabla_k (E_{ck} - E_{vk}) \right|} \tag{9.58}$$

At that point, the conduction and VBs are parallel, having larger values (singularities) at energies. In the dielectric function, the structure is given by these larger values (Wooten and Weaire 1984).

9.4.5 BAND-TO-BAND TRANSITION

The dielectric function and optical constants of semiconductors are distributed throughout a vast array of photon energies defined by transitions between different bands of band structures. At small frequencies, the mechanism of electronic conduction phase of semiconductors is linked with the free carriers. As the energy of phonon increases and then eventually becomes equivalent to the energy gap, a new phase of conduction may occur. Then an electron may excite due to phonon from the filled state in the VB to an empty state in the CB. This process is called interband transition. In this process, the photon is absorbed and a hole is left behind due to the formation of the excited electronic state. This is termed a quantum mechanical process (Schneider 2004).

Band-to-band transitions that are optically influenced are resonance transitions and these transitions are linked to band structure by the JDOS and matrix elements. For the near band-gap transitions, the optical transition theory between conduction and VBs can be generalized by appropriate mass estimation, supposing measurable absorption expressions as a function of light energy and vertical band shapes. Strong direct transitions and weak indirect transitions take place at the edge of the band while depending on the behavior of the CB. In fact, the involvement of forbidden transitions leads to further change in the absorption transitions away from the band edge. Direct or indirect transitions can be unequivocally defined based on the slope of the absorption spectrum in the intrinsic spectrum near the edge of the band, and, with the support of an applied magnetic field, the effective mass can be estimated from the time of electromagnetism of the near-edge absorption due to the Landau separation of the band structures (Philipp and Ehrenreich 1963).

In semiconductors, the transition can be made between two different energy states via destroying or creating photons during the process. In addition, transitions are optically active between the CB and the VB (interband transitions), since the low CB typically consists of s-like bands, whereas p-like states are at the top VB. Between VB states and CB states, absorption spontaneous emission, and stimulated emission can all occur.

9.4.6 Interband Transitions in Semiconductors

Interband transitions in semiconductors usually occur at frequencies above where the contribution of free charge carrier becomes important. For the whole dielectric constant, the equation will be

$$\varepsilon = \varepsilon_{core} + \frac{4\pi i}{\omega} \left[\sigma_{Drude} + \sigma_{interband} \right] \tag{9.59}$$

ε_{core} covers the involvement of all methods that are not measured clearly in the previous equation; this would contain both interband and intra-band transitions as these processes are important in studies of electronic properties of semiconductors (Huldt 1971).

In semiconductors, the thermal excitation of free carriers takes place at room temperature because of the small bandgap. Thus, the absorption of considerable free charge carrier at room temperature occurs through doping or thermal excitation. In infrared (IR) and visible regions, interband transition takes place. For example, in germanium, there is direct transition with relation to optical absorption. The direct bandgap in semiconductors means the lowest point or minima of the CB in k space that occurs right above the extreme of the VB (Bai et al. 2004).

While in indirect semiconductors, the minimum energy is shifted by k vector in the CB concerning the top of the VB. As the photon has small momentum, so when the momentum of the crystal is conserved, then both initial and final band states are permitted. These transitions are represented by vertical lines called vertical transitions. The momentum provided by phonons or impurities cause non-vertical transitions. Transitions, including impurity or phonon, and photon state are characterized indirectly. The examples of indirect transitions are interband transition near zone boundary or from top of the valence to bottom of the CB and the other is an intra-valley transition that is responsible for carrier absorption (Cardona and Yu 2011).

From initial to final state transition is given as

$$\omega_{f \leftarrow i} = \frac{\Gamma_i \hbar}{\left(E_f - E_i \right)^2 + \Gamma_i^2} \left| \sum_a \frac{\langle f | H | a \rangle \langle a | H | i \rangle}{E_i - E_a + i\Gamma_i} \right|^2 \tag{9.60}$$

The E_i and E_f are energies of initial and final states and Γ_i is energy of relaxation according to second-order perturbation theory ($H = H_{\hbar\omega} + H_{ep}$). One transition route is given by $H_{\hbar\omega}$, optical transition induces from initial state $| i \rangle$ to intermediate state $| a \rangle$. For the overall conservation of momentum and energy, the H_{ep}, i.e., the phonon contribution, completes the transition by taking it from initial state $| a \rangle$ to final state $| f \rangle$.

On the other hand, the first step can be done by the phonon perturbation and the second step by the optical perturbation. Here, we are not studying the less probable two-phonon (H_{ep} active in both steps) or two-photon developments ($H_{\hbar\omega}$ active in both steps) (Dresselhaus 2001).

Following are the important factors in interband transitions:

- In bandgap energy, we considered interband transitions to get threshold energy. To exhibit a threshold for allowed electronic transition because of an interband transition, we considered the dependence of frequency of real part of conductivity (ω).

- These transitions have been either direct in which there is conservation of momentum of crystal $\vec{k}$: $E_v(\vec{k}) \rightarrow E_c(\vec{k})$ or indirect in which a phonon is included as a vector for conduction and VBs and vary by the wave vector of phonon. The conservation momentum of crystals gives $\vec{k}_{valence} = \vec{k}_{conduction} \pm \vec{q}_{phonon}$. In case of considering direct transitions, some may ask about the momentum conservation of crystal with respect to a photon.
- The transition in semiconductors depends on the coupling between conduction and VBs, and this is determined by the magnitude of the elements of the momentum matrix coupling the state of the VB v and the state of the CB c: $\left|\langle v|\vec{p}|c \rangle\right|^2$. This value of dependence is calculated using Fermi's Golden rule and perturbation interaction due to the electromagnetic field with electrons in solids.
- The interband transitions occur due to the Pauli exclusion principle from the filled state just below the Fermi level to an empty state above the Fermi level.

If the energy gap between the two bands is approximately constant for several $\vec{k}$ values, then the photons with given energy are more efficient in generating a bandgap transition. In this case, there are several primary and final states that can be linked by identical photon energy. This is probably clearer to see that if we permit a photon to have narrow bandwidth, then that narrow bandwidth will be useful over several values of k if $E_v(\vec{k}) \rightarrow E_c(\vec{k})$ and does not rapidly change with $\vec{k}$. From Figure 9.1, it can be seen that for optical source, the largest contribution per unit bandwidth formed around $\vec{k} = 0$. This is because the study of the energy band structure is important for optical measurements of semiconductors. Thus, it is supposed that interband transitions are most vital for near-band extrema values; band extrema are highlighted by the optical structure, which provides data about the energy bands at definite points of the Brillouin zone (Xia, Chang, and Li 2004).

9.4.7 Types of Interband Transitions

Interband transitions are categorized as direct or vertical or indirect or non-vertical near the basic absorption edge. Comparison to the momentum of a k-vector state just at boundary of the Brillouin field, the momentum of light $(\hbar k = \hbar n \omega / c)$ is insignificant. Thus, electrons with a specified wave vector in a range could only make transitions to bands in high states having exactly the same wave vector due to momentum conservation. These transitions are referred to as vertical transitions. Non-vertical transitions can occur only with the support of phonons or more entities that help reserve momentum (Bhattacharya, Fornari, and Kamimura 2011).

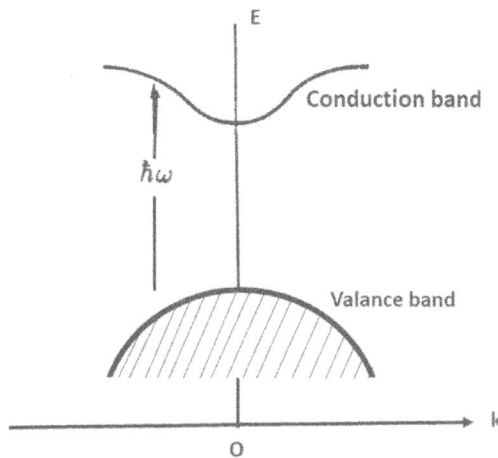

FIGURE 9.1   Schematic illustration of permitted interband transitions.

The absorption coefficient of direct interband transition depends on photon energy and band structure. Direct transitions can be divided into forbidden and allowed transitions, which depend on dipole matrix elements between bands or the coefficient of absorption vanishes or is finite in the first approximation. Semiconductors such as Si, Ge and GaP have indirect bandgaps where minimum energy of CB and maximum energy of VB do not take place at identical k value (Cardona and Pollak 1966).

The electron should not establish a direct transition from the higher state of the VB to the lower state of the CB, since this would invalidate the momentum conservation. Such transitions can still occur, and even as a two-step phase that requires another particle to cooperate and then can be established by the principle of second-order perturbation. The most commonly observed particle is energy $\hbar\Omega_k$ interval phonon that can be either produced or absorbed during the transition. In some situations, it is important to consider elastic dispersion processes due to conduction electrons or dislocations; they are less common than interactions with phonons. The photon provides the energy needed, while the required momentum is supplied by the phonon. The probability of transition depends on the elements of the electron matrix and the electronic states, such as those in the direct case, and also on the temperature-dependent phonon-electron interaction (Alonso et al. 2001).

In the case of absorption, the components for optical absorption include an arriving photon, an electron-occupied energy state of the VB and a vacant energy state of the CB. The VB is usually complete in equilibrium and the CB has been mostly empty, so these specifications are readily met. There was only optical absorption from two states of an atom with a small range of photon energies (the line width of the transition). By comparison, optical absorption can happen over a wide spectrum of photon energies in a semiconductor as well as the energy of the photon is approximately equal to the energy of the bandgap. Since a hole in the valance band and an electron in the conductive band are created by the absorption process, it is often called optical generation. By releasing photons, electrons generated at higher energies than the bandgap can readily relax to the lower CB energy levels (band edge levels) (Kalt and Klingshirn 2019a).

Additionally, holes created deeper in the VB would "float up" to the edge of the VB. The process of spontaneous emission in semiconductors requires a vacant VB and an occupied CB energy state. This is the inverse of the typical equilibrium condition, but there will be a limited number of complete states in the CB and vacant states in the VB at a finite temperature. It is also possible to generate electrons and holes through optical absorption (Schmitt-Rink, Miller, and Chemla 1987).

The process of spontaneous emission just like with absorption is possible for the top of bandgap for a broad range of photons. However, in practice, the states of the CB are probably full for the states with the lowest energies. Therefore, it is most likely that the spontaneous emission detected from a semiconductor sample has photon energy that is almost equivalent to the bandgap energy. This is also called spontaneous optical recombination as it destroys holes and electrons. Therefore, ordinary silicon is not convenient for light emission as silicon has an indirect bandgap, so the lowest energy state in the CB and maximum energy state in the VB have dissimilar momentum. The optical transition of indirect and direct interbands varies differently and depends upon photon energy.

9.5   REFRACTIVE INDEX

Even though the characteristics of electromagnetic waves in a semiconductor crystal are controlled by Maxwell's equations, it is generally not simple to precisely calculate the electric field correlated with an electromagnetic wave. Most electromagnetic wave sensors measure the intensity of electromagnetic wave proportional to $E(t)^2$. In addition, most sensors do not respond quickly to calculate the electromagnetic wave's time variation and the time-averaged intensity (Abbar et al. 2001; Kuno 2008).

The information of time-averaged intensity is indirectly measured by frequency of wave where time variation measures through Fourier transformation. The dielectric function of any crystal can be measured by common techniques such as refractive index, transmission coefficient and reflection coefficient. If the crystal is intensely absorbing, then it may turn out to be difficult to calculate the coefficient of transmission and refractive index. The most typical methods used to determine a crystal's dielectric function include measurements of coefficient of reflection, refractive index and transmission coefficient. If the crystal is heavily absorbing, then the calculations of the coefficient of transmission and refractive index become difficult. In such situations, the dielectric function alone can be calculated from the reflectivity (coefficient of reflection) (Herve and Vandamme 1994).

The refractive index of the crystal is determined by using the law of refraction. The refractive index is equal to c/v, and in nonmagnetic crystals $\varepsilon = (n)^2$, here, ε is tensor, n is not. We cannot explicitly obtain the transformed values of n on rotating the coordinate axis. Instead, the transformed value of ε must first be obtained and the new values of n must then be determined. The refractive index for certain crystal symmetry can be expressed by an ellipse recognized as the indicatrix. The indicatrix is the ellipse described by when the x, y, z coordinate axes are selected to be along the major axis of the dielectric tensor.

$$\frac{x^2}{n_x^2} + \frac{y^2}{n_y^2} + \frac{z^2}{n_z^2} = 1 \tag{9.61}$$

Here, the refractive indices are n_x, n_y and n_z, which equal $\sqrt{\epsilon_x}$, $\sqrt{\epsilon_y}$ and $\sqrt{\epsilon_z}$, respectively.

This equation for cubic solids lessens to sphere so that n does not depend on direction of **E**. In the most common situation, for the given wave vector k, one can measure the refractive indices values for polarized light by passing the normal plane to k through the origin. By changing the path of light over crystal and calculating the values of refractive indices for mutually perpendicular light, the crystal's dielectric tensor could be determined (Balzaretti and Da Jornada 1996).

The dielectric function ε of linear regime and susceptibility χ can be defined by the following relations:

From Equation (9.51)

$$\begin{aligned} \mathbf{D} &= \varepsilon_0 \mathbf{E} + \mathbf{P} \\ \mathbf{D} &= \varepsilon_0 (1 + \chi)\mathbf{E} \\ \mathbf{D} &= \varepsilon \mathbf{E} = (\varepsilon_1 + i\varepsilon_2)\mathbf{E} \end{aligned} \tag{9.62}$$

where P, E and D, are polarization filed, electric field of free space and polarization, displacement field inside semiconductor, respectively: χ and ε are dimensionless quantities and ϵ_0 is the free space permittivity, all these quantities can describe all optical properties of semiconductors. The refractive index of any material is linked with permittivity ε as next:

From Equation (9.41),

$$N = \sqrt{\varepsilon} = n + ik$$

Here, n and k are real and imaginary parts of refractive index, N, that are also stated as optical constants and represent linear optical properties of semiconductors. The absorption component of optical energy is presented by imaginary part k (Abraham and Firth 1990).

In the spectral states where absorptive procedures are absent or weak, as in the region of the sub bandgap range, **k** has very low value; however, in states of strong absorption, the level of **k** is large. The difference in the real part is generally minimal, e.g., at room temperature in GaAs, the values of real parts in visible regions ranging from 1.4 eV to 6 eV, and the values of imaginary part ranging from 10^{-3} eV below bandgap at 1.41 eV to extreme of 4.1. In contrast, refractive index remains almost constant in region of gap ranging from 3.61 to 3.8 at 1.4 eV to 1.9 eV with the values of 5.1 at 2.88 eV and 1.26 at 6 eV, respectively (Ravindra, Ganapathy, and Choi 2007)

9.6 REFLECTION, REFRACTION AND TRANSMISSION

While crossing the interface between any two media, light can be reflected or refracted. In this regard, boundary conditions for magnetic and electric fields at interface are of prime importance along with laws of refraction and reflection describing momentum variation and frequency conservation. All these terms are briefly discussed next.

9.6.1 BOUNDARY CONDITIONS AND LAWS OF REFRACTION AND REFLECTION

Consider an electric field $\mathbf{E}$ of incident beam, which is in parallel polarization to the plane of incidence defined by wave vector $\mathbf{k}$. Boundary conditions imply that electric field and wave vectors of reflected and transmitted beams occur in the same plane, while perpendicular to them are magnetic field. Following all the boundary conditions in the situation, we get

$$D_n^I = D_n^{II} \tag{9.63}$$

which shows continuity of normal component of along the interface. Similarly,

$$B_n^I = B_n^{II} \tag{9.64}$$

Consequently, the tangential components of $\mathbf{E}$ and $\mathbf{H}$ are given as follow.

$$E_t^I = E_t^{II} \tag{9.65}$$

$$H_t^I = H_t^{II} \tag{9.66}$$

Equations (9.63–9.66) reveal the boundary conditions for magnetic and electric fields, which enables one to calculate properties of reflected, refracted and transmitted beams for a given incident wave (Haught et al. 1999; Römer 2009).

It is a fact that light has the same frequency in all three beams, which is presented by Equation (9.67). According to classical point of view, atoms with frequency ω perform forced oscillations under the effect of incident field, and radiation also occurs at the same frequency ω. This is also proved by quantum point of view, considering conservation law of energy along with the fact that every photon having energy ω can only be transmitted or reflected back.

$$\omega_i = \omega_r = \omega_{tr} \tag{9.67}$$

Laws of reflection and Snell's law of refraction are also important in the aspect of reflection and refraction studies. According to law of reflection,

$$\alpha_i = \alpha_r \tag{9.68}$$

For k_i, k_r and e_n occurring in the same plane and e_n as the normal to the interface, Snell's law can be followed:

$$\frac{\sin \alpha_i}{\sin \alpha_{tr}} = \frac{n_{II}}{n_I} \tag{9.69}$$

In the other case, refraction from optically thick to thin part of the medium, critical angle α_i^c is considered for $\alpha_{tr} = 90°$. The previous equation is rewritten as

$$\alpha_i^c = \arcsin \frac{n_{II}}{n_I} \tag{9.70}$$

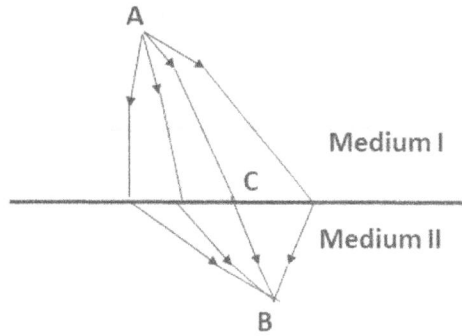

FIGURE 9.2 Various possible paths of a light beam traveling from A to B.

In case of angle of incidence greater than critical angle, the entire beam reflects back and no transmission at all. The abovementioned boundary conditions usually require finite amplitudes in the second medium, which shows a momentary wave in the medium propagating parallel with the surface. In direction perpendicular to the interface, the field amplitude exponentially decreases, which is elaborated in the following. The reflected wave possesses the same intensity as that of incident one, and the process is known as total internal reflection. In the second case, where medium has thickness comparable to the wavelength of the beam, and material II is covered with medium I, this gives rise to a propagating transmitted wave. As a result, reflected wave intensity decreases and the case is known as attenuated total reflection (ATR) (Lüders 2017).

The above-deduced laws of refraction and reflection can be described by Fermat's principle. According to this principle, the optical path length, which is the product of index of refraction n and geometric path length, between any two given points is usually minimum; this is shown in Figure 9.2 in the case of refraction. Light chooses the smallest possible path from all the available paths to reach the final position, and the equation for this is presented by

$$\delta \int n ds = \delta \left(n_I \overline{AC} + n_{II} \overline{CB} \right) = 0 \tag{9.71}$$

Variation of optical path length δ disappears in case of selected path; thus, laws of reflection and refraction can be deduced from Equation (9.71) (Kalt and Klingshirn 2019b).

9.6.2 Transmission and Reflection through Parallel Plane Matter Slab

Considering a parallel plane slab of matter through which a beam passes, we have to elaborate the transmission and reflection through thickness d and smooth surfaces. Properties of incident light such as its coherence length l_c, polarization also effect reflection $\hat{R}$ and transmission $\hat{T}$ in addition to incident angle and medium II, where medium I is usually considered vacuum or air (Meschede 2007; Saleh and Teich 2019). Reflected beam is easily detected in case of reflection from the front face. In visible part with wavelength of $\lambda \cong 500nm \ll d$, transmitted beam is most likely to be attenuated, with optical density $\alpha(\omega)d \gg 1$. Thin samples have different optical properties from the bulk matter and thus make it difficult to calculate transmission from it. The same is the case with reflection $\hat{R}$ in the case of $\alpha(\omega)d \gg 1$, though $\alpha(\omega)$ may be quite small. $\hat{R}$ reduces to R in case of normal incidence.

$$\hat{R}(\omega) = R(\omega) \text{ for } \alpha(\omega)d \gg 1 \tag{9.72}$$

The range of $\alpha(\omega)d$ is usually considered between 1 and 5 for convenience. Reflection occurs mainly at the front and back surfaces due to low chances of multiple reflections because of absorption.

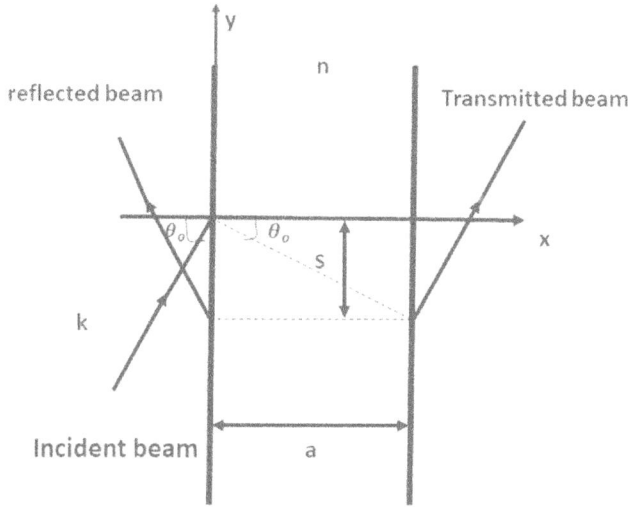

FIGURE 9.3 Multiple reflections of a beam through parallel plane matter slab.

Multiple reflections of a beam are presented in Figure 9.3. Transmission coefficient found in this state is written next:

$$\hat{T}(\omega) = \left[1 - R^{I \to II}(\omega, \alpha)\right] \exp\left[-\alpha(\omega) d \cos^{-1} \alpha_{tr}\right]\left[1 - R^{II \to I}(\omega, \alpha_{tr})\right] \quad (9.73)$$

This equation then simplifies to Equation (9.74) in the case of normal incidence for low absorbent material considering the condition $R^{I \to II} = R^{II \to I} = R(\omega)$.

$$\hat{T}(\omega) \cong \left[1 - R(\omega)\right]^2 e^{-\alpha(\omega)d} \quad (9.74)$$

Taking into account Equation (9.72), relation of reflection coefficient is as follows:

$$\hat{R}(\omega) \cong R(\omega) + \left[1 - R(\omega)\right]^2 e^{-2\alpha(\omega)d} \cong R(\omega) \quad (9.75)$$

Properties of parallel plane slab are influenced by multiple reflections in case of medium with optical density $\alpha(\omega)d \leq 1$. Relation between coherence length l_c and optical path greatly affects the performance. In case of short coherent length, $l_c^{-1} \ll d$, we must consider normal incident intensities in previously discussed case, which lead to the following modifications in the expression (Kalt and Klingshirn 2019b; Thelen 1989).

$$\hat{T}(\omega) \cong \frac{\left[1 - R(\omega)\right]^2 e^{-\alpha(\omega)d}}{1 - R^2(\omega) e^{-2\alpha(\omega)d}} \cong \left[1 - R(\omega)\right]^2 e^{-\alpha(\omega)d} \text{ for } \alpha d \lesssim 1 \quad (9.76)$$

9.7 OPTICAL PROPERTIES OF INTRINSIC EXCITONS

Optical properties of insulators and semiconductors are usually explained by considering phonon excitations in IR, while plasmon takes part in the case of metals from IR to ultraviolet (UV). Excitons play a part in determining optical properties of bulk semiconductors around the bandgap for UV and IR regions, and while in case of insulators, they are valuable in UV range only. To study optical properties of intrinsic exciton in semiconductors, band-to-band transition, forbidden transition in direct and indirect bandgap semiconductors and interaction transitions are the main points of focus (Kalt and Klingshirn 2019b).

9.7.1 Direct Transitions in Exciton

Strong coupling of exciton with the field of radiation in semiconductors for direct band-to-band transitions can lead to many modifications in optical properties. As a consequence, signatures of exciton in optical spectra get changed, which are to be discussed later. Mixed states of electromagnetic radiation and excitation give rise to exciton-polariton. In direct transition, the phenomenon of an electron-hole pair is generated by an incident photon. Exciton is formed by the Coulomb interaction among the holes and electrons. The vertical lines represent these excitons or the photon exchange among the electrons and holes (Haug and Koch 2009).

One photon absorption coefficient among final and initial states is written as

$$\alpha(\omega) \propto \left|\langle f|H_D|i\rangle\right|^2 \delta(E_i - E_f + \hbar\omega) \tag{9.77}$$

While the refractive index from $\alpha(\omega)$ by Kronig transformation is as given next ((Kalt and Klingshirn 2019b; Peyghambarian, Koch, and Mysyrowicz 1993)):

$$n^2(\omega) - 1 \propto \sum_{z \neq i} \frac{\langle i|H_D|z\rangle\langle z|H_D|i\rangle}{E_z - E_i - \hbar\omega} \tag{9.78}$$

9.7.1.1 Reflection Spectra

Flat and structureless reflection spectra are obtained in direct band-to-band transitions. As the lattice temperature is increased, scattering of phonons increases and thus giving rise to broadening of excitons resonance. Samples with high impurity content and low crystallinity can experience washing out of excitons' resonance even at low temperature. For low temperature range, the reflection spectra in direct bandgap semiconductors usually occur at bandgap by exciton polariton. Influence of temperature on reflectivity of CdS has been already studied in Bohnert, Schmieder, and Klingshirn (1980), Hümmer (1973) and Varshni (1967).

Some uniaxial, hexagonal materials such as ZnO, ZnTe and CdS have mixed mode polariton. The reflectivity spectra of ZnTe and CdS with varying thickness of exciton-free layer are reviewed in Kalt and Klingshirn (2019b) and Maier, Schmieder, and Klingshirn (1983).

9.7.1.2 Absorption Spectra

In resonance region, the absorption coefficient occurs in the range of $10^4 - 10^6\,cm^{-1}$, and hence it is difficult to measure absorption spectra quantitatively. If the thickness of samples is in the range of micrometer, intensity of transmitted light goes to zero. By increasing temperature, homogeneous broadening due to scattering from excited phonons can be increased, but it becomes difficult to observe finer details of the absorption spectra. Exciton improvement of photon energies above the bandgap considerably boosts the absorption even at room temperature (Liebler, Schmitt-Rink, and Haug 1985). Effect of temperature on the absorption explains the shifting of exciton peak according to the bandgap. The reason behind this is the interaction of optical phonons with the exciton. Absorption is given by (Kalt and Klingshirn 2019b; Klingshirn 2010):

$$\alpha(\hbar\omega) = \alpha_o \exp\left[-\frac{\sigma(T)(E_o - \hbar\omega)}{K_B T}\right] \text{ for } \hbar\omega < E_o \tag{9.79}$$

9.7.2 Forbidden Transitions in Exciton

Dipole-forbidden transitions or spin flips can be the reason for low oscillating strength of the exciton because they involve a third element such as a phonon for its creation. These forbidden transitions are studied separately for direct and indirect band semiconductors next.

9.7.2.1 Transitions in Direct Gap Semiconductors

Though semiconductors with direct bandgap can possess band-to-band or dipole-allowed transitions but there also can occur some forbidden transitions giving rise to exciton states with low oscillator strength. Longitudinal exciton, spin flip and forbidden dipole transition symmetries occur due to the presence of these spin triplet states. When $n \geq 2$ and $L = 1,2,$ there is a rise to exciton states in single-photon transition as magnetic dipole or electric quadrupole states. Example of this kind of weak transition is CdS. In the case of $n = 1$, dipole-allowed transitions occur for $E \perp c$ polarization and $E \parallel c$, only weak exciton transitions are possible to occur. In this case, aperture angle of light beam creates deviation in $k \perp c$, which leads to small oscillating strength (Kalt and Klingshirn 2019b; Shinada and Sugano 1966).

Another situation of dipole forbidden transition may occur in some kind of semiconductors, which may possess direct bandgap but due to the same parity; transition among lower CB and upper VB is not allowed. Some important materials in this regard are SnO_2, Cu_2O, GeO_2 and TiO_2 that possess dipole forbidden transition in $L = 0$ state and weak oscillator strength at $L = 1$ state due to odd parity (Hayashi and Katsuki 1952).

9.7.2.2 Transitions in Indirect Gap Semiconductors

Some important semiconductors like Si, Ge, AgBr and GaP have indirect gap. In these materials, lowest states of free excitons don't occur at $K = 0$, and these states are unable to couple with radiation field due to conservation of momentum. The third particle like a phonon is required to make sure momentum conservation in both emission and absorption practices. This momentum conservation condition is not a necessary condition for certain alloys like $Al_{1-x}Ga_xAs$ above transition temperature. At high temperature, light quanta are absorbed with both phonon absorption and emission, while at low temperature, photon is absorbed only in the emission of phonons (Kalt 1996). In indirect gap materials, exciton states don't possess sharp peaks due to a whole density of states, making absorption spectra somewhat simpler with the following approximations:

$$\alpha_{cm}(\hbar\omega) \propto \alpha_0 \left[\hbar\omega - E_{ex}(k_0) + \hbar\omega_{ph} \right]^{\frac{1}{2}} \left(1 + N_{ph} \right) \tag{9.80}$$

$$\alpha_{abs}(\hbar\omega) \propto \alpha_0 \left[\hbar\omega - E_{ex}(k_0) - \hbar\omega_{ph} \right]^{\frac{1}{2}} N_{ph} \tag{9.81}$$

Here, α_0 represents matrix elements of transition, which weakly depend upon phonon momentum, while $E_{ex}(k_0)$ represents minimum energy of excitons and phonon energy as shown by $\hbar\omega_{ph}$ and phonon numbers by N_{ph}. Due to the presence of phonons in absorption process in indirect gap materials, transition probability decreases by several orders of magnitude in comparison with the direct bandgap of semiconductors. Absorption coefficient for indirect transition occurs in the range of 1 cm^{-1} to 10^2 cm^{-1}. Due to such low value of absorption coefficient, no particular change occurs in the reflection spectra. At low temperature for $k_BT < \hbar\omega_{ph}$, absorption process under phonons becomes feebler. On increasing energy up to a specific level for indirect excitons, some kind of direct excitons can occur in the absorption spectra for some materials like Ge. Rapid decay occurs for direct exciton, causing strong damping, thus making it difficult to observe fine structures for higher states (Würfel, Finkbeiner, and Daub 1995).

The third particle involvement makes the luminescence yield smaller for indirect materials as compared to direct gap semiconductors. Due to this reason, bulk silicon can't be utilized in LED. But the other aspect is that it makes the lifetime of exciton of indirect materials longer in the range of 1 to many microseconds based on materials purity, while lifetime of exciton of direct materials occurs in nanoseconds range (Kalt and Klingshirn 2019b).

9.7.3 INTRAEXCITON TRANSITION

Intraexcitonic transitions are the transitions occurring between exciton sublevels for indirect gap materials like Si or Ge. Continuous wave laser pumping is utilized and transitions occur from 1s to 2p states due to carrier spins interactions and angular momentum by applying 10 meV–12 meV. The transition energies do not coincide with the absorption of phonons, and excitons' lifetime is considered longer due to high exciton density than bound exciton density. Cu_2O also has dipole forbidden transition apart from its direct gap, which induces longer lifetime in it and transition occurs from 1s para exciton to 2p states. Defects in the states, overtones of phonons or biphonons cannot be neglected in such kind of intraexciton transitions (Nikitine 1986; Timusk et al. 1978).

9.8 EFFECT OF ELECTRIC FIELD ON EXCITONS AND CARRIERS

Excitons in semiconductors are greatly affected by applying electric field **E**, which is commonly termed stark effect that introduces exciton resonance shifting and splitting in both linear and quadratic states. Variation in electric field introduces band-to-band transitions, which is known as Franz-Keldysh effect. As the static E is applied, which induces band tilting and allows tunneling of wave function of electrons and holes to the forbidden gap, this creates a redshift in absorption spectra and thus affects index of refraction in a transparent range below the resonance of excitons. This has been explained by Kramer-Kronig relation (Madelung and Mollwo 1971; Pankove 1975).

9.8.1 APPLICATION OF ELECTRIC FIELD ON BULK SEMICONDUCTORS

In bulk samples, exploration of stark effect is a difficult task due to some specific requirements. First, absorption band spectral width must be comparable to the energy of **E** at low temperature such as,

$$eE_s a_B \geq \Delta_{LT} \tag{9.82}$$

For this condition to be met, field must be of the order of $10^6\,Vm^{-1}$, which may vary for different materials. However, it may disturb exciton resonance or broaden the field due to field ionization or impact ionization. Field ionization occurs due to tunneling of finite Coulomb barrier at finite fields. The other effects may occur due to the gain of energies by carriers at finite temperature, which may ionize excitons creating more carriers and broadening of exciton resonance. This is called impact ionization. Quadratic stark effect on excitons results in shifting and broadening of excitons in the absorption band (Nelson et al. 2001).

Energies of minima and maxima in band-to-band transitions follow the following relation (Kalt and Klingshirn 2019b):

$$\left(E_n - E_g\right)^{3/2} \sim n \tag{9.83}$$

9.8.2 APPLICATION OF ELECTRIC FIELD ON LOW-DIMENSIONAL STRUCTURE

In bulk semiconductors, problem of field ionization can be controlled by restraining holes and electrons in barriers like in quantum wires, wells or dots. As **E** increases, which is in parallel orientation with the layers, holes and electrons shift toward the potential sides resulting in the decrease of energy separation, along with redshift in resonance. There is no restriction on $\Delta n_z = 0$, and E_s mixes states irrespective of odd or even parity, thus giving rise to forbidden transitions for $E_s = 0$ (Miller et al. 1985).

Electric field applied to superlattice planes causes shift in adjacent walls and change in energy is recognized as

$$\Delta E = e|E_s|.d \tag{9.84}$$

Here, energy is represented by E_s and superlattice period by d. Due to this, mini-band formation can't be originated and a Wannier-Stark ladder is produced. This quantum stark effect is suitable for low-dimensional structures like quantum wires or dots or quantum islands. Charges that are confined in the locality of quantum dots, their electric fields produce fluorescence intermittency of single dots (Henneberger, Schmitt-Rink, and Göbel 1993).

9.9   EFFECT OF MAGNETIC FIELD ON CARRIERS AND EXCITONS

To study effect of magnetic field on excitons, two natural energy scales known as Rydberg and cyclotron energy must be comprehended. Cyclotron energy is $\hbar\omega_c = \hbar\left(\frac{eB}{\mu}\right)$.

The weak field limit has the regime,

$$R_y^* \gg \hbar\omega_c \Rightarrow \gamma = \hbar\left(\frac{eB}{\mu R_y^*}\right) \ll 1 \tag{9.85}$$

where R_y^* is Rydberg energy and μ is reduced exciton mass.

While the strong field limit has the following condition:

$$R_y^* \ll \hbar\omega_c \Rightarrow \gamma \gg 1 \tag{9.86}$$

Application of magnetic field on semiconductors results in Zeeman splitting, diamagnetic shift and introduction of landaus levels, which are to be elaborated.

In the absence of magnetic field, exciton states possess a non-vanishing magnetic moment, which upon the application of magnetic field, align relative to **B**, along with Zeeman splitting. Spin of holes or electrons contributes in creating magnetic moment. In case of singlet or triplet states with spin S, energy difference is written as

$$\Delta E_z = \pm\frac{1}{2}\left|g_e \pm g_h\right|\mu_B \mathbf{B} \tag{9.87}$$

Electrons and holes g factor can vary from value of 2 because of crystal symmetry or band structure. When n exceeds from 2, orbital quantum number involvement introduces additional factors in magnetic moment, which also depends upon magnetic quantum number (Kalt and Klingshirn 2019b).

9.9.1   APPLICATION OF MAGNETIC FIELD ON NONMAGNETIC BULK SEMICONDUCTORS

In bulk semiconductors, introduction of magnetic field creates Zeeman splitting and diamagnetic shift in their magnetic optic spectra, which has been demonstrated by example of CdS in the study (Blattner et al. 1982). Upon applying magnetic field, dipole-allowed exciton resonance occurs at $n_B = 1$, along with Zeeman splitting while no diamagnetic shift is observed in it.

The previous example demonstrates Zeeman effect for ground state, while bulk ZnSe possesses exciton resonance for $n_B = 2$ and $l = 1$. In this case, fine splitting can be observed in the absence of magnetic field, while increase in magnetic strength results in the increment in splitting along with diamagnetic shift due to a large span of relative motion of holes and electrons. Some other magneto-optics materials that display a similar behavior include Cu-halides, normal semiconductors like GaAs and some kind of insulating alkali halides. Cu_2O displays dipole forbidden excitons due to collaboration of para and ortho excitons, whereas CdS and layered GaSe show magneto-stark effect (Klingshirn 2012a).

Landau states are introduced at the highest value of **B**, which allow determination of electrons and holes masses. Some semiconductors (like GaAs) that have small exciton binding energies, their states shift in continuum states due to application of magnetic field and make Fano interference lineshapes (Bar-Ad et al. 1997).

9.9.2 APPLICATION OF MAGNETIC FIELD ON BULK MAGNETIC SEMICONDUCTORS

Some magnetic semiconductors that have magnetic ions are known as diluted or semi-magnetic semiconductors (DMS or SMSC). Magnetic moments when couple with excitons or carriers strongly influence their optical properties. II-VI or III-V semiconductors are doped with magnetic ions to form their higher concentration alloys that are then called DMS. Their bandgaps, phonons, localization and binding energies decide their particular compositions. Most widely used dopants are Mn^{+2}, Fe, Ni, Co and Gd that have spin and magnetic moments (Ando 2000). In DMS, coupling of magnetic ions is usually antiferromagnetic, and there is zero contribution toward ferro- or paramagnetism. Paramagnetic or ferromagnetic behavior increases with x_{nom} through a maximum and vanishes when antiferromagnetic domains take over them. In DMS, sp-d exchange occurs, in which transition takes place in s or p orbital of host's conduction and valance bands. While 3d orbital of magnetic ions participate in transition, magnetic moments are greatly polarized by any stray or bound carrier, resulting in massive Zeeman splitting and huge g-factor of excitons.

9.10 CONCLUSION

Semiconductors and their linear effects are very important for their implementation in technical applications. In this chapter, we have tried to review major linear optical properties and effects appeared in semiconductors. Basic elementary theory of solids based on famous Maxwell's equations is reviewed in detail. Maxwell's equations entail all the information about light, so behavior of light in vacuum and matter is devoured with Maxwell's equations. Behavior and modification of Maxwell's equations have been considered interrogated in order to understand the nature of interaction between light and matter to derive the linear response coefficients. Linear response coefficients are dielectric function and complex refractive index. Review of linear wave equation also gave information about attenuation and absorption coefficient. We have examined theoretically the band-to-band transition as a property of linear semiconductors. We have also studied the factors necessary for interband transitions. Band-to-band transitions that are optically influenced are resonance transitions and these transitions are linked to band structure by the JDOS and matrix elements. Strong direct transitions and weak indirect transitions take place at the edge of the band while depending on the behavior of the CB. The optical transition of indirect and direct interband varies differently and depends on photon energy, and the absorption coefficient of direct interband transition depends on photon energy and band structure. The crystal's refractive index is defined by the law of refraction, and if the crystal is highly absorbing, then for calculation of refractive index, the reflectivity has to be calculated first. The laws of reflection and refraction along with transmittance coefficient suggest that wave vectors while crossing materials interface follow momentum conservation and Fresnel's formula. Additionally, Fermat's principle is also significant in discussing properties of light waves and follows depiction of light extermination by scattering and absorption. Dipole-allowed band-to-band transition along with forbidden and intraexciton transitions in semiconductors is also studied. Dipole-allowed excitons have the highest oscillator strength, while forbidden transition has low strength. In forbidden transition, some excitons may follow dipole-allowed transitions, which are studied in direct and indirect gap materials. Effects of external magnetic and electric fields are studied on bulk and low-dimensional structures, which explain new trends in optical behavior of semiconductors utilized in numerous applications. In addition to these external fields, there are certain internal factors like strain fields and piezoelectric effects.

REFERENCES

Abbar, B, B Bouhafs, H Aourag, G Nouet, and P Ruterana. 2001. "First-principles calculations of optical properties of AlN, GaN, and InN compounds under hydrostatic pressure." *Physica Status Solidi (B)* 228 (2):457–460.

Abraham, NB, and WJ Firth. 1990. "Overview of transverse effects in nonlinear-optical systems." *JOSA B* 7 (6):951–962.

Alonso, MI, and M Garriga. 2018. "Optical properties of semiconductors." In *Spectroscopic ellipsometry for photovoltaics*, 89–113: Springer.

Alonso, MI, K Wakita, J Pascual, M Garriga, and N Yamamoto. 2001. "Optical functions and electronic structure of CuInSe2, CuGaSe2, CuInS2, and CuGaS2." *Physical Review B* 63 (7):075203.

Ando, K. 2000. "Magneto-optics of diluted magnetic semiconductors: new materials and applications." In *Magneto-optics*, 211–244: Springer.

Aspnes, DE, and SD Yoo. 1999. "High-resolution spectroscopy with reciprocal-space analysis." *Physica Status Solidi (B)* 215 (1):715–723.

Bai, L, Z Lin, Z Wang, C Chen, and M-H Lee. 2004. "Mechanism of linear and nonlinear optical effects of chalcopyrite AgGaX2 (X = S, Se, and Te) crystals." *The Journal of Chemical Physics* 120 (18):8772–8778.

Balzaretti, NM, and JAH Da Jornada. 1996. "Pressure dependence of the refractive index of diamond, cubic silicon carbide and cubic boron nitride." *Solid State Communications* 99 (12):943–948.

Band, YB. 2006. *Light and matter: electromagnetism, optics, spectroscopy and lasers.* Vol. 1: John Wiley & Sons.

Bar-Ad, S, P Kner, MV Marquezini, S Mukamel, and DS Chemla. 1997. "Quantum confined Fano interference." *Physical Review Letters* 78 (7):1363.

Bhattacharya, P, R Fornari, and H Kamimura. 2011. *Comprehensive semiconductor science and technology*: Newnes.

Blattner, G, Gl Kurtze, G Schmieder, and C Klingshirn. 1982. "Influence of magnetic fields up to 20 T on excitons and polaritons in CdS and ZnO." *Physical Review B* 25 (12):7413.

Bohnert, K, G Schmieder, and C Klingshirn. 1980. "Gain and reflection spectroscopy and the present understanding of the electron–hole plasma in II–VI compounds." *Physica Status Solidi (B)* 98 (1):175–188.

Cardona, M, and FH Pollak. 1966. "Energy-band structure of germanium and silicon: the k·p method." *Physical Review* 142 (2):530.

Cardona, M, and PY Yu. 2011. "A chapter in volume 2 of comprehensive semiconductor science and technology." In *Optical properties of semiconductors*, edited by H Kamimura: Elsevier.

Cohen, ML, and JR Chelikowsky. 2012. *Electronic structure and optical properties of semiconductors.* Vol. 75: Springer Science & Business Media.

Dean, KJ. 1984. "Waves and fields in optoelectronics: Prentice-Hall series in solid state physical electronics." *Physics Bulletin* 35 (8):339.

Dietl, T. 1994. "Magnetic semiconductors." *Interaction* 2 (5):9.

Dresselhaus, MS. 2001. "Solid state physics. Part II. Optical properties of solids." *Lecture Notes (Massachusetts Institute of Technology, Cambridge, MA)* 17:15–16.

Faber, KT, and KJ Malloy. 1992. *The mechanical properties of semiconductors.* Vol. 37: Academic Press.

Fox, M. 2001. *Optical properties of solids*: Oxford University Press.

Furdyna, JK. 1988. "Diluted magnetic semiconductors." *Journal of Applied Physics* 64 (4):R29–R64.

Glazer, M, G Burns, and AN Glazer. 2012. *Space groups for solid state scientists*: Elsevier.

Gu, L, V Srot, W Sigle, C Koch, P van Aken, F Scholz, SB Thapa, C Kirchner, M Jetter, and M Rühle. 2007. "Band-gap measurements of direct and indirect semiconductors using monochromated electrons." *Physical Review B* 75 (19):195214.

Haug, H, and SW Koch. 2009. *Quantum theory of the optical and electronic properties of semiconductors*: World Scientific Publishing Company.

Haught, A, RT Walls, JD Laney, A Leavell, and SS, Patricia. 1999. "Child and adolescent knowledge and attitudes about older adults across time and states." *Educational Gerontology* 25 (6):501–517.

Hayashi, M, and K Katsuki. 1952. "Hydrogen-like absorption spectrum of cuprous oxide." *Journal of the Physical Society of Japan* 7 (6):599–603.

Henneberger, F, S Schmitt-Rink, and EO Göbel. 1993. *Optics of semiconductor nanostructures*: Vch Pub.

Herve, P, and LKJ Vandamme. 1994. "General relation between refractive index and energy gap in semiconductors." *Infrared Physics & Technology* 35 (4):609–615.

Huldt, L. 1971. "Band-to-band Auger recombination in indirect gap semiconductors." *Physica Status Solidi (A)* 8 (1):173–187.

Hümmer, K. 1973. "Interband magnetoreflection of ZnO." *Physica Status Solidi (B)* 56 (1):249–260.

Ibach, H, and H Lueth. 2009. *Solid-state physics. An introduction to principles of materials science*: Springer Science & Business Media.

Kalt, H. 1996. "Many-body effects in multi-valley scenarios." In *Optical properties of III–V semiconductors*, 41–124: Springer.

Kalt, H. 2004. "CdTe quantum wells." In *Optical properties. Part 2*, 13–48: Springer.

Kalt, H, and CF Klingshirn. 2019a. *Semiconductor optics 1*: Springer.

Kalt, H, and CF Klingshirn. 2019b. *Semiconductor optics 1: linear optical properties of semiconductors*: Springer Nature.

Kittel, C, P McEuen, and P McEuen. 1996. *Introduction to solid state physics*. Vol. 8: Wiley.

Klingshirn, C. 2010. "Intrinsic linear optical properties close to the fundamental absorption edge." In *Zinc oxide*, 121–168: Springer.

Klingshirn, CF. 2012a. "Excitons under the influence of (external) fields." In *Semiconductor optics*, 423–455: Springer.

Klingshirn, CF. 2012b. *Semiconductor optics*: Springer Science & Business Media.

Köhler, A, and H Bässler. 2015. *Electronic processes in organic semiconductors: an introduction*: John Wiley & Sons.

Kuno, M. 2008. "An overview of solution-based semiconductor nanowires: synthesis and optical studies." *Physical Chemistry Chemical Physics* 10 (5):620–639.

Liebler, JG, S Schmitt-Rink, and H Haug. 1985. "Theory of the absorption tail of Wannier excitons in polar semiconductors." *Journal of Luminescence* 34 (1–2):1–7.

Lüders, K. 2017. *Pohl's introduction to physics*: Springer.

Madelung, OW, and E Mollwo. 1971. "Elektrotransmission und Pockelseffekt an Zinkoxid." *Zeitschrift für Physik A Hadrons and nuclei* 249 (1):12–30.

Maier, W, G Schmieder, and C Klingshirn. 1983. "Magnetopolaritons in ZnTe." *Zeitschrift für Physik B Condensed Matter* 50 (3):193–208.

Maxwell, JC. 1865. "VIII. A dynamical theory of the electromagnetic field." *Philosophical Transactions of the Royal Society of London* 155:459–512.

Meschede D. 2007. "Light rays." *Optics, Light and Lasers*:1–27.

Miller, DAB, DS Chemla, TC Damen, AC Gossard, W Wiegmann, TH Wood, and CA Burrus. 1985. "Electric field dependence of optical absorption near the band gap of quantum-well structures." *Physical Review B* 32 (2):1043.

Nelson, D, B Gil, MA Jacobson, VD Kagan, N Grandjean, B Beaumont, J Massies, and P Gibart. 2001. "Impact ionization of excitons in an electric field in GaN." *Journal of Physics: Condensed Matter* 13 (32):7043.

Nikitine, S. 1986. "On the possibility of observation and intensity of (nn′) and (nn) transitions between exciton states in Cu₂O." *The Journal of Physics and Chemistry of Solids* 47 (1):115.

Pankove, JI. 1975. *Optical processes in semiconductors*: Courier Corporation.

Pendry, JB. 2000. "Negative refraction makes a perfect lens." *Physical Review Letters* 85 (18):3966.

Peter, YU, and M Cardona. 2010. *Fundamentals of semiconductors: physics and materials properties*: Springer Science & Business Media.

Peyghambarian, N, SW Koch, and A Mysyrowicz. 1993. *Introduction to semiconductor optics*: Prentice-Hall, Inc.

Philipp, HR, and H Ehrenreich. 1963. "Optical properties of semiconductors." *Physical Review* 129 (4):1550.

Probert, M. 2011. "Electronic structure: basic theory and practical methods." In *Scope: graduate level textbook. Level: theoretical materials scientists/condensed matter physicists/computational chemists*, edited by RM Martin: Taylor & Francis.

Ravindra, NM, P Ganapathy, and J Choi. 2007. "Energy gap–refractive index relations in semiconductors – An overview." *Infrared Physics & Technology* 50 (1):21–29.

Rizwan, M, I Iqra, SSA Gillani, I Zeba, M Shakil, and Z Usman. 2021. "First-principles investigation of structural, electronic, and optical response of SnZrO₃ with Al inclusion for optoelectronic applications." *Physics of the Solid State* 63 (1):134–140.

Römer, H. 2009. *Theoretical optics: an introduction*: Wiley-VCH.

Saleh, BEA, and MC Teich. 2019. *Fundamentals of photonics*: John Wiley & sons.

Schmitt-Rink, SDABM, DAB Miller, and DS Chemla. 1987. "Theory of the linear and nonlinear optical properties of semiconductor microcrystallites." *Physical Review B* 35 (15):8113.

Schmitt-Rink, S, DS Chemla, and DAB Miller. 1989. "Linear and nonlinear optical properties of semiconductor quantum wells." *Advances in Physics* 38 (2):89–188.

Schneider, T. 2004. *Nonlinear optics in telecommunications*: Springer Science & Business Media.

Schwoerer, M, and HC Wolf. 2007. *Organic molecular solids*: John Wiley & Sons.

Shinada, M, and S Sugano. 1966. "Interband optical transitions in extremely anisotropic semiconductors. I. Bound and unbound exciton absorption." *Journal of the Physical Society of Japan* 21 (10):1936–1946.

Shindo, K, A Morita, and H Kamimura. 1965. "Spin-orbit coupling in ionic crystals with zincblende and wurtzite structures." *Journal of the Physical Society of Japan* 20 (11):2054–2059.

Smith, DR, JB Pendry, and MCK Wiltshire. 2004. "Metamaterials and negative refractive index." *Science* 305 (5685):788–792.

Stern, F. 1963. "Elementary theory of the optical properties of solids." *Solid State Physics* 15:299–408.

Stöcker, H. 1999. *Taschenbuch mathematischer Formeln und moderner Verfahren*: Deutsch.

Stößel, W. 2013. *Fourieroptik: Eine Einführung*: Springer-Verlag.

Stratton, JA. 2007. *Electromagnetic theory*. Vol. 33: John Wiley & Sons.

Thelen, A. 1989. *Design of optical interference coatings*: McGraw-Hill Companies.

Timusk, T, H Navarro, NO Lipari, and M Altarelli. 1978. "Far-infrared absorption by excitons in silicon." *Solid State Communications* 25 (4):217–219.

Trügler, A. 2016. *Optical properties of metallic nanoparticles: basic principles and simulation*. Vol. 232: Springer.

Van Vechten, JA, and SP Keller. 1980. *Handbook on semiconductors*, 1111. Vol. 3: North-Holland.

Varshni, YP. 1967. "Temperature dependence of the energy gap in semiconductors." *Physica* 34 (1):149–154.

Voon, LCLY, and M Willatzen. 2009. *The k-p method: electronic properties of semiconductors*: Springer Science & Business Media.

Wooten, F, and D Weaire. 1984. "Generation of random network models with periodic boundary conditions." *Journal of Non-Crystalline Solids* 64 (3):325–334.

Würfel, P, S Finkbeiner, and E Daub. 1995. "Generalized Planck's radiation law for luminescence via indirect transitions." *Applied Physics A* 60 (1):67–70.

Xia, J-B, K Chang, and S-S Li. 2004. "Electronic structure and optical property of semiconductor nanocrystallites." *Computational Materials Science* 30 (3–4):274–277.

Yuan, L-D, H-X Deng, S-S Li, S-H Wei, and J-W Luo. 2018. "Unified theory of direct or indirect band-gap nature of conventional semiconductors." *Physical Review B* 98 (24):245203.

Zhang, Z, W Zhang, F Wang, LM Tolbert, and B J Blalock. 2012. "Analysis of the switching speed limitation of wide band-gap devices in a phase-leg configuration." In *2012 IEEE Energy Conversion Congress and Exposition (ECCE)*.

中山隆史. 1989. "ML Cohen and JR Chelikowsky, Electronic Structure and Optical Properties of Semiconductors, Springer-Verlag, Berlin and Heidelberg, 1988, xii+ 264p., 23.5× 16 cm, 9,241 円 (Springer Series in Solid-State Sciences, 75)[大学院向教科書, 専門書]." 日本物理学会誌 44 (9):690.

10 Computational Techniques on Optical Properties of Metal-Oxide Semiconductors

Jhilmil Swapnalin, Prasun Banerjee, Chetana Sabbanahalli, Dinesh Rangappa, and Kiran Kumar Kondamareddy

CONTENTS

10.1 INTRODUCTION

Massive advancement in the theoretical study of materials using cheap computational features has led material scientists to precisely predict materials' behavior and nature, even without conducting actual experiments.

Optical computing is one of the best computing technologies since the 1980s. However, initially, optical computing was applied to very few materials. The material limitations made optical computing trapped in 1980. With the discovery of new materials with wide bandgap oxides, the researchers could make the best use of semiconductor optical computing to obtain various properties shown in Figure 10.1 [1].

Ultrafast computers with high efficiency and performance quality are mandatory with this rapid growth in computing technology. Optical computing has been known for many years now. In the 1980s, optical computing was a central topic of research, but there was a steep fall in the domain due to material limitations. However, this area of optical computing is growing again with the improved nonlinear organic material and semiconductor technology [2].

Electronic structure theory has a major difficulty in computing the optical properties of semiconductors using the first-principle theoretical tool. Its reason is the interaction between electrons and interaction between conduction band electrons with valence band holes. This problem was addressed by various approaches like the Bethe-Salpeter equation (BSE) in electron-hole interaction and GW approximation in electron self-energy [3,4]. In a fictitious Kohn-Sham (KS) system, the density-density response is targeted as a function by TDDFT or time-dependent density functional theory. Various approaches like local-density approximation (LDA) [5, 6], orbital-dependent functionals [7,8], and TDDFT were implemented to get an accurate bandgap in the KS system but could not compute the optical properties. At the same time, the standard method adiabatic

DOI: 10.1201/9781003188582-10

155

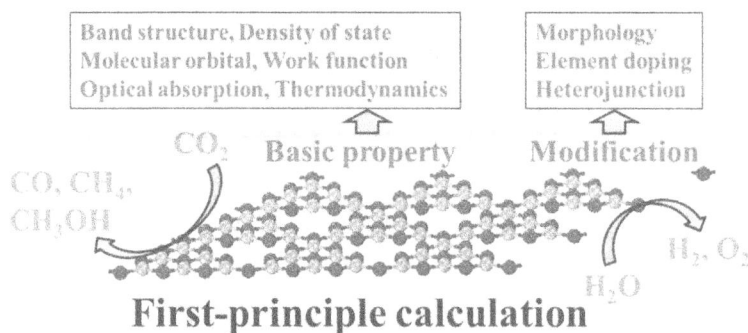

FIGURE 10.1 Semiconductor optical computing to obtain various properties [1].

LDA (ALDA) [9] and its extension generalized-gradient approximation (GGA) and exact exchange approach failed significantly. In the recent decade, a new approximate functional in ground-state DFT calculates the correct KS bandgap [10, 11]. It is known as meta-GGA (MGGA), which focuses on density and its gradient and noninteracting kinetic energy density [12]. Its adiabatic approximation leads to enhanced optical property study of semiconductors and provides "ultranonlocality," thus being a better substitution of TDDFT [13].

Furthermore, the nanoquanta kernel [14–17] derived from the BSE kernel (four-point) under TDDFT shows high accuracy but is much more expensive than the kernels as mentioned earlier. Finally, the jellium-with-gap model (JGM) proposes a nonempirical static xc kernel [18] suitable for determining qualitative and quantitative insights of semiconductors. This model is an extension of the ALDA kernel and goes well with experimental data. This density functional kernel is an ab initio approach able to estimate good Eg. This kernel, independent of frequency, was found to describe both bound excitons in ionic insulators and excitonic effects in semiconductors. Moreover, it provided absorption spectra as per the experimental results [19].

Bandgap evaluation is also appropriately done using mBJ or modified Becke-Johnson functional, consistent with experiments, where GGA and mBJ potential evaluate the exchange-correlation potential [20]. First-principle calculations by hybrid Heyd-Scuseria-Ernzerhof (HSE06) method are used in inorganic perovskite semiconductor optical computing, thereby confirming band type [21].

10.2 METAL-OXIDE SEMICONDUCTORS AND COMPUTATIONAL METHODS

10.2.1 Zinc Oxide

Zinc oxide (ZnO), with a bandgap of 3.37 eV, is one of the most extensively studied under a wide bandgap oxide semiconductor category. It helps make transparent conducting oxide films, short-wavelength light-emerging devices, and surface acoustic wave devices. By controlled addition of impurities, the electrical and optical properties of ZnO can be varied dramatically. The application of zinc oxide can be in various sectors, including piezoelectric transducers, chemical sensors, spin functional devices, optical waveguides, biomedical applications, and specific target applications of ZnO nanostructures that made zinc oxide, a material with huge demand [22].

ZnO, a wide bandgap semiconducting eco-friendly material with the possible synthesis techniques, turns out to be the most used material for optical computing. In recent years, zinc oxide, which is synthesized at different thermal conditions with different dopants, is obtained in different shapes, which results in various physical properties. Researchers had obtained zinc oxide in rods, wires, sheets, nanofillers, tetrapods, and nanoribbons/nanobelts. With all these modified synthesis techniques, their electronic and optical properties have been enhanced, resulting in the best emission and absorption conditions. Zinc oxide nanoparticles show different optical and electronic properties compared to bulk zinc oxide material, which brings in the interest to study quantum confinement effects in the nano-environment. Researchers have reported that the zinc oxide with a particle size of 0.7- to 1.5-nm radii possesses a bandgap of ~3.7–3.8 eV.

In contrast, the normal bandgap value of zinc oxide is ~3.3–3.4 eV at increased particle size. This particle size variation also leads to the different emission properties showing different colors at different spectrum regions. For example, the bulk zinc oxide with the bandgap of 2.4–2.1 eV shows emission near the green-yellow region, while 2.7–2.8 eV shows emission in the blue region, and 1.8–1.9 eV emits the orange spectrum. The observed visible color-changing properties of zinc oxide based on bandgap explain the luminous properties. Also, this visible luminous property is crucial in explaining the absorption strength of the material [23].

10.2.1.1 DFT Techniques

DFT uses fundamental laws of quantum mechanics, a vastly employed theory that accurately computes the atomic and molecular structures, which helps understand the calculable properties of a material.

Generally, electronic structure is computed by finding the approximate solutions to the Schrödinger equation applied to an external potential comprising N-interacting electrons. Nevertheless, in this approach, even minimal values of N result in complicated wave functions, and the computation effort increases very rapidly with the increasing value of N, leading to significant difficulty in describing the larger systems, whereas, in DFT, instead of N wave functions, the fundamental variable is one-body density. Because of this, the density n(r) becomes a function of three spatial coordinates, making DFT an efficient tool to do electronic structure computation of large systems. As a result, DFT is now commonly employed to study the electronic structure of semiconductors [24].

With DFT, the morphological and defective properties of zinc oxide have been studied; through this, it is identified that the vacancy of oxygen at some points clearly shows the green luminescence properties. Few researchers also claim that in bulk zinc oxide materials, the green laminations are observed due to the low bandgap values. Using LDA and GGA helps in understanding the better DFT properties of zinc oxide. The DFT simulation of zinc oxide bulk materials predicted the bandgap as 0.8 eV. In contrast, the theoretical bandgap of zinc oxide is 3.4 eV with this difference in bandgap values; it is very inaccurate to study optical properties of zinc oxide. TDDFT simulation helps us understand the electron-hole interaction and exchange-correlation of zinc oxide and help us report the appropriate bandgap of the material with not many complications. The TDDFT includes geometrical relaxation and polarizability of the material with these approximations; the computational methodology expresses the bulk materials initially very small, and then for nanoparticles, it increases. The absorption lamination and optical behaviors of ZnO are studied in the 1D, 2D, and 3D structures of ZnO.

TDDFT simulation has been carried out on 1D, 2D, and 3D zinc oxide bulk and nanoparticles. The 1D and 2D zinc oxide nanomaterials do not show much variation, and the properties of luminescence are not much observed when compared with electronic and optical properties, whereas 3D zinc oxide nanomaterials have been showing vital luminescence properties experimentally, and the theoretical TDDFT calculations have also been showing the large visible bandgap for 3D materials.

When 3D nanostructures of zinc oxide are simulated through TDDFT, researchers have drastically observed the missing oxygen atoms in the zinc oxide structure. This oxygen vacancy in the structure is responsible for the zinc oxide visible emission bandgap region. Both neutral oxygen atoms at the ground state can be vacant oxygen for opposite spin electrons, which leads to both singlet and triplet states of the optimizer zinc oxide, and by DFT simulation, it is also observed that the geometrical structure of zinc oxide is varied with the oxygen vacant atoms, leading to the shrinkage of the distance between Zn and Zn. The oxygen vacancy in the structure clearly shows that the oxygen atom is in a neutral state that makes the distance between 2 Zn molecules very less, and then the atoms start moving from singlet state to triplet state. The displacement of electrons in the defect state is shown in Figure 10.2, which shows the Zn triangle [23].

The experimental and theoretical studies were carried out on doped zinc oxide Nano and bulk materials. It was observed that zinc oxide doped semiconductor in its lattice parameters and

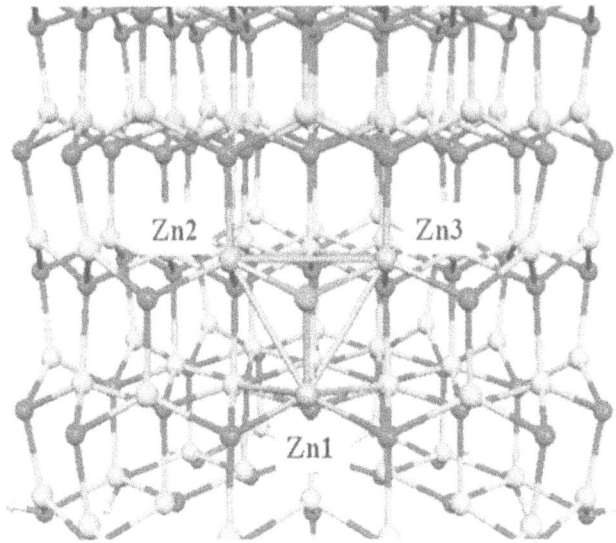

FIGURE 10.2 The displacement of electrons in the defect state, which shows the Zn triangle [23].

distortion is observed with this the razor change in the energy band gap, but the bandgap was extremely less varied when compared to that of the experimental band gap of doped zinc oxide material. Through this we can conclude that the doping does not affect much on zinc oxide and the bandgap value approximately remain same. Considering tin(Sn) as a dopant for zinc oxide experiments was carried out when Sn was added to the zinc oxide material; there was not much change in its lattice parameter when the concentration of Sn was increased, then there is a slight distortion in a lattice parameter, which results in a variation of the bandgap. The varied bandgap of tin-doped zinc oxide and the theoretical value of zinc oxide were comparatively similar to not many changes; through this, it is concluded that dopant concentration will not affect the bandgap of zinc oxide [22].

10.2.2 VANADIUM OXIDE

The vanadium oxide is a prime focus for its insulator-metal transition (IMT), which occurs in the lower oxides like VO, V_2O_3, and VO_2 [25–27]. Vanadium oxides, in general, are chromogenic that alter their optical properties due to external factors like photo-radiation and temperature change. At the same time, there is a discontinuity during the IMT [28].

Vanadium dioxide undergoes first-order transition, namely IMT, usually induced via thermal [29], electrical [30], optical means [31], or simply through strain [32]. This phase transition gives rise to carrier-density changes across the IMT [33], which marks significant variation in the optical properties [34]. As a result, advanced optical properties are exploited for various applications like optical limiting [35, 36], switching [37, 38], nonlinear isolation [39], and thermal-emission engineering [40].

Vanadium oxides help make optoelectronic devices like smart windows [41], resistive random-access memories [42], and thermal sensors [43]. Moreover, the compounds showing IMT can be manipulated through pressure and temperature, which changes the structural, optical, and electronic properties [44]. Hence, vanadium oxides are mostly preferred to study both experimentally and theoretically due to this phase transition.

Out of all oxide forms, V_2O_5 tends to be more stable with high anisotropic electrons and optical properties. Though V_2O_5 has been studied more experimentally, theoretical study is minimum due to its orthorhombic structure and complicated block building.

In the case of VO_2, this phase change can be instantaneous or linear depending on the VO_2 thin film characteristics. It depends on the quality, grain size, amount of potential impurities, and degree of the strain [45–47]. A variety of optical and optoelectronic devices can be derived using this material, and its computational design can be analyzed by using the data of complex refractive index across IMT. Experimental results were verified using effective-medium-theory (EMT), which is implemented to study the gradually changing optical properties throughout the phase transition of IMT and do optical simulations [48].

In the 1990s, theoretical calculations of optical properties of metal oxides gained attention; the approach was first-principle calculations—the optical computing of the single-crystal V_2O_5 performed through spectroscopic ellipsometry. Experimental results were verified using wave functions of the first principle by band-structure calculations performed through an orthogonalized linear combination of atomic orbitals (OLCAO). V_2O_5 is recorded as a semiconductor having an indirect bandgap of about 2.0 eV. Due to its conspicuous mixture of two localized bands, it is usually considered semiconductors exhibit a large energy gap of ~3.3 eV. The energy assignment of the significant spectral features documented theoretically showed good agreement with experimental spectra when local-density-functional theory (LDFT) was applied. However, intensity discrepancies were noted, which is due to the high anisotropy of V_2O_5, while the single crystals of VO_2 being less anisotropic show better agreement when optical properties are compared [49].

In V_2O_3, the vanadium sesquioxide, the optical properties were depicted using the combination of LDA and dynamical mean field approximation (DMFT) [50] by considering e^π_g and a_{1g} carrier transitions under DMFT [51, 52]. Theoretical computation was advanced by considering the experimental variations in lattice parameters as a function of temperature [53]. When LDA+DMFT calculation at two temperatures 300 and 700K is merged with documented real data of lattice parameters, a shrink in LDA bandwidth was observed. The optical conductivity from DMFT shows a prominent pseudogap as predicted from experiments. Optical conductivity calculated using LDA+DMFT is shown in Figure 10.3 [51].

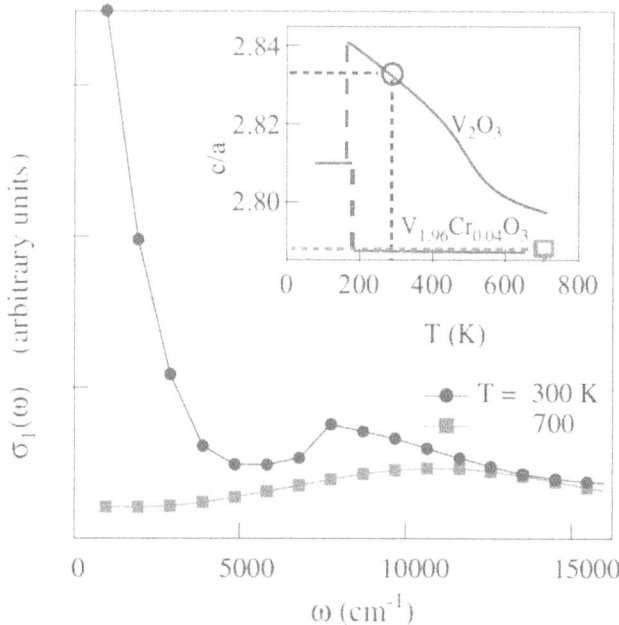

FIGURE 10.3 Optical conductivity calculated using LDA+DMFT [51].

Some theoretical calculations showed the structure of VO_2 could be appropriately evaluated with the cluster extension of DMFT combining DFT [54] and verified the optical properties by LDA+DMFT [55], showing good agreement with experimental results.

When the transmission of VO_2 film synthesized by spray pyrolysis was studied, it was around 63% for a wavelength of 2000 nm [56, 57], the energy bandgap Eg value of being 2.5 eV. Moreover, the extinction coefficient (k) and refractive index (n) of the film decreased as the wavelength increased.

When the coherent potential approximation (CPA) and the first-principle theory of Korringa-Kohn-Rostoker (KKR) methods were implemented to VO_2, the stability, electrical, and optical properties were well examined [58].

Later, to investigate the electronic as well as the optical properties of the VO_2 (monoclinic), first-principle calculations with TB-mBJ and SCAN+U approaches were taken. Many facts like Eg value, band splitting, anisotropic behavior, and various inter- and intra-band optical responses coincided with experimental data. Ultimately it is shown how the Mott-Hubbard electron correlation is the primary inducer for MIT in VO_2 to verify electronic and optical characteristics [59].

To compute the optical properties, two methods were used, namely GGA+U with the HSE06 (Heyd-Scuseria-Ernzerhof hybrid functional). From the spectra complex, optical conductivity and complex dielectric function were derived, showing the GGA+U approach describes the optical properties of vanadium oxides precisely [60]. Furthermore, the optical conductivity of various oxides is plotted and compared between theoretical and experimental data, as shown in Figure 10.4 [60–65].

10.2.3 INDIUM TIN OXIDE

Out of various semiconductors, In_2O_3: Sn or ITO (indium tin oxide) is a degenerate n-type semiconductor [66]. With plasma frequency in the range of near-infrared, it exhibits an optical bandgap of ~3.6 eV [66–69]. Due to these properties, the metal oxide is usually transparent and reflective in the visible and IR spectral range, respectively. The conductive nature of ITO favors it for various applications like the thin film form that helps make electrodes, gas sensors, heat shields, photoelectrolysis, solar cells, heterogeneous catalysis, and also flat panel displays [66–74]. Thin films prepared by different synthesis and annealing methods determine the electronic and optical potential of ITO [67–77]. In this metal oxide, In_2O_3 is taken as a substrate material, and controlled doping of Sn is done and obtained through high transmittance with low resistivity as seen in the visible part of the spectral range [78].

In electrochemical studies, ITO surfaces have been modified to alter the properties of electronic and optical factors. For a comprehensive examination of the material, many theoretical and experimental interpretations have been performed. From the theoretical point of view, electronic and optical properties were investigated using DFT based on a model, which consists of the entire cubic unit cell of parent indium oxide.

The computational study of the material aims mainly to verify whether the electronic and optical properties coincide with the experimental findings or not [69–79]. Theoretical calculations showed how minimal change in the doping amount of tin changes the charge carrier concentration, affecting the optical and electronic properties. It has helped determine the optical bandgap, effective free electron mass, charge carrier concentration, conduction band orbital nature, and plasma frequency in the metal oxide. DFT calculations with quantum chemical software DMol3 and the Perdew and Wang GGA functionals were done to calculate band structure at Fermi option (0K) and the rest other calculations at thermal option (~6300K), as shown in Figure 10.5 [80].

P-type $InSnO_2$ is prepared by Ji et al. by two different methods, spray pyrolysis and sol-gel dip-coating, which show that the indium proportion determines structure and conduction modes and has no effect on optical properties.

For further investigation using first-principle calculations employing a plane wave basis having periodic boundary conditions, Cambridge Serial Total Energy Package (CASTEP) code [81] was implemented. Perdew-Bruke-Eruzerhof+ GGA [82] described e^-–e^- exchange and correlation effects under DFT, thus giving proper perception on structural, electronic, and optical properties

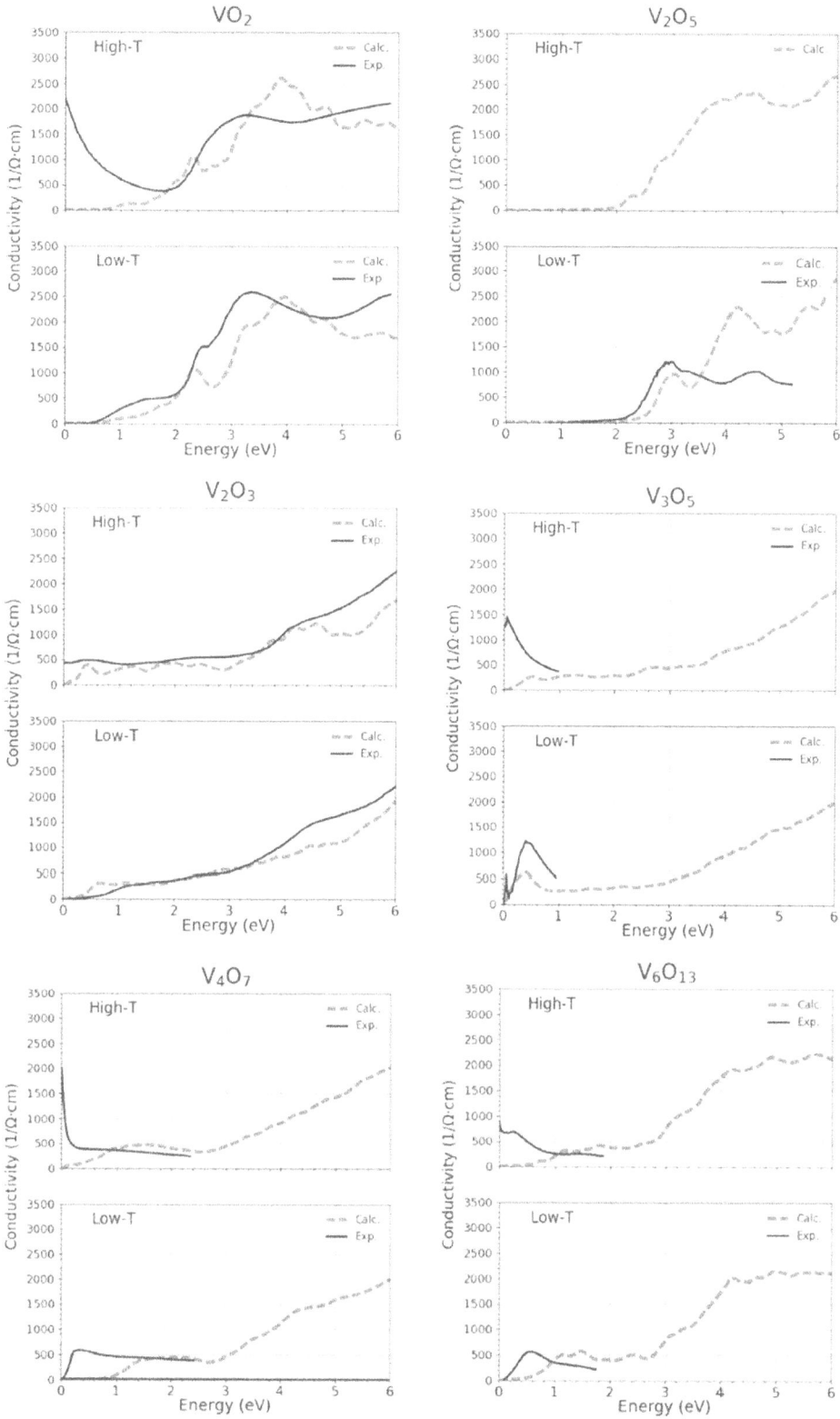

FIGURE 10.4 Optical conductivity of various oxides is plotted and compared between theoretical and experimental data [60].

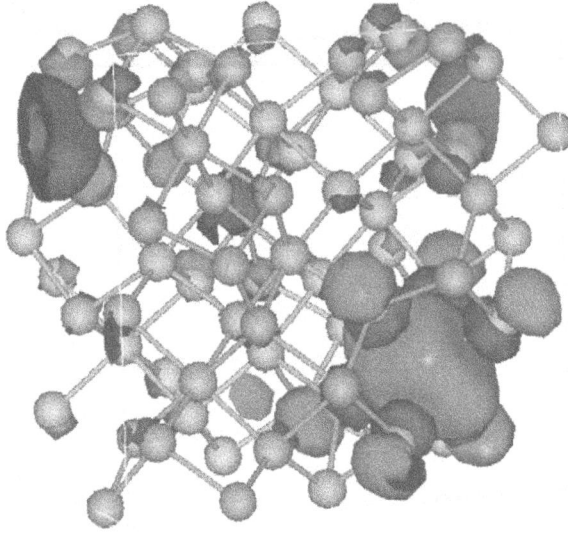

FIGURE 10.5   Conduction band of ITO orbitals calculated with DFT techniques [80].

of the metal oxide. The reliable option to obtain the optical properties is the usage of pseudo-potentials (Norm-conserving) [83].

From the dielectric function, optical parameters such as absorption and reflectivity are calculated, while from the Kramers-Kronig relation, the imaginary dielectric function is obtained from the real part. Small changes in both conduction band and optical bandgap have zero effect on P-type $InSnO_2$ as seen in the UV region, and a surge in the values of reflectivity, absorption, and dielectric function in the IR region is seen due to the exciton effect and transition to empty bands from occupied ones [84].

To mitigate discrepancies reported in previous studies of electronic, structural, and optical properties in ITO with notable differences between experimental and theoretical studies, corrections in standard DFT are made. Semiempirical DFT þ U in the domain of standard DFT was used to carry out the first-principles study for ITO. On-site corrections within LDA þ U enhance theoretical outcomes and mitigate the differences when both experimental and theoretical data are compared. Actually, correction is applied to the cation 4d orbitals, which in turn reproduces hybridized energy levels under LDA þ U, which ultimately enhances the optical spectral calculation [85].

Optical absorption of the semiconductor is also studied through LDA by using the on-site Coulomb interaction parameter U (LDA + LDA þ U), correcting the valence band levels, leading to improvement in the theoretical bandgap coincident with the experimental value [86].

Moreover, in the case of the ITO ultrathin films, DFT with a nonequilibrium Green function approach was taken to theoretically note the structural and optical properties at different temperatures and thicknesses. It was noted that with the increase in thickness, absorption coefficient and transmittance decreased, whereas reflectivity increased with nearly similar optical spectra [87].

10.3   CONCLUSION

In summary, we studied semiconductor optical computing in metal-oxide semiconductors. First-principle calculations based on DFT and modified approaches in its domain were also applied to study the properties. The study's main aim was to enhance the optical properties of the semiconductors like zinc oxide, vanadium oxide, and ITO using theoretical methods and compare the simulated outcomes with the experimental results already available. TDDFT was favorable for zinc oxide to calculate bandgaps accurately. Vanadium oxides known for their IMT are studied using different theoretical methods and approximations like EMT, OLCAO, LDFT, LDA+DMFT, for which spectral features, band type, and value are estimated. In ITO, the optical study was done primarily by DFT, and corrections in the domain using LDA þ U were found to be improving theoretical predictions. In fact, due to thorough optical computing using theoretical simulations, the advantages of the semiconductors in making various optoelectronic devices were confirmed.

ACKNOWLEDGMENTS

One of us, P. Banerjee, received TARE fellowship from SERB, India, with TAR/2021/000032 research grant.

REFERENCES

1. Zhu, Bicheng, Liuyang Zhang, Bei Cheng, and Jiaguo Yu. "First-principle calculation study of tri-s-triazine-based g-C3N4: a review." *Applied Catalysis B* 224 (2018): 983–999.
2. Goswami, Debabrata. "Optical computing." *Resonance* 8, no. 6 (2003): 56–71.
3. Hybertsen, Mark S., and Steven G. Louie. "Electron correlation in semiconductors and insulators: band gaps and quasiparticle energies." *Physical Review B* 34, no. 8 (1986): 5390.
4. Albrecht, Stefan, Lucia Reining, Rodolfo Del Sole, and Giovanni Onida. "Ab initio calculation of excitonic effects in the optical spectra of semiconductors." *Physical Review Letters* 80, no. 20 (1998): 4510.
5. Gross, E. K. U., J. F. Dobson, and M. Petersilka. "Density functional theory of time-dependent phenomena." *Density Functional Theory II*, 181 (1996): 81–172.
6. Li, Jingbo, and Lin-Wang Wang. "Band-structure-corrected local density approximation study of semiconductor quantum dots and wires." *Physical Review B* 72, no. 12 (2005): 125325.
7. Kümmel, Stephan, and Leeor Kronik. "Orbital-dependent density functionals: theory and applications." *Reviews of Modern Physics* 80, no. 1 (2008): 3.
8. Görling, Andreas. "Orbital-and state-dependent functionals in density-functional theory." *The Journal of Chemical Physics* 123, no. 6 (2005): 062203.
9. Gross, E. K. U., and Walter Kohn. "Local density-functional theory of frequency-dependent linear response." *Physical Review Letters* 55, no. 26 (1985): 2850.
10. Becke, Axel D., and Erin R. Johnson. "A simple effective potential for exchange." (2006): 221101.
11. Tran, Fabien, and Peter Blaha. "Accurate band gaps of semiconductors and insulators with a semilocal exchange-correlation potential." *Physical Review Letters* 102, no. 22 (2009): 226401.
12. Ullrich, Carsten A., and Zeng-hui Yang. "Excitons in time-dependent density-functional theory." *Density-Functional Methods for Excited States*, 368 (2014): 185–217.
13. Nazarov, V. U., and G. Vignale. "Optics of semiconductors from meta-generalized-gradient-approximation-based time-dependent density-functional theory." *Physical Review Letters* 107, no. 21 (2011): 216402.
14. Sottile, Francesco, Valerio Olevano, and Lucia Reining. "Parameter-free calculation of response functions in time-dependent density-functional theory." *Physical Review Letters* 91, no. 5 (2003): 056402.
15. Marini, Andrea, Rodolfo Del Sole, and Angel Rubio. "Bound excitons in time-dependent density-functional theory: optical and energy-loss spectra." *Physical Review Letters* 91, no. 25 (2003): 256402.
16. Stubner, R., I. V. Tokatly, and O. Pankratov. "Excitonic effects in time-dependent density-functional theory: an analytically solvable model." *Physical Review B* 70, no. 24 (2004): 245119.
17. von Barth, Ulf, Nils Erik Dahlen, Robert van Leeuwen, and Gianluca Stefanucci. "Conserving approximations in time-dependent density functional theory." *Physical Review B* 72, no. 23 (2005): 235109.
18. Tsolakidis, Argyrios, Eric L. Shirley, and Richard M. Martin. "Effect of coupling of forward-and backward-going electron-hole pairs on the static local-field factor of jellium." *Physical Review B* 69, no. 3 (2004): 035104.
19. Trevisanutto, Paolo E., Aleksandrs Terentjevs, Lucian A. Constantin, Valerio Olevano, and Fabio Della Sala. "Optical spectra of solids obtained by time-dependent density functional theory with the jellium-with-gap-model exchange-correlation kernel." *Physical Review B* 87, no. 20 (2013): 205143.

20. Mahmood, Q., and M. Hassan. "Systematic first principle study of physical properties of $Cd_{0.75}Ti_{0.25}Z$ (Z= S, Se, Te) magnetic semiconductors using mBJ functional." *Journal of Alloys and Compounds* 704 (2017): 659–675.

21. Rasukkannu, Murugesan, Dhayalan Velauthapillai, and Ponniah Vajeeston. "A first-principle study of the electronic, mechanical and optical properties of inorganic perovskite Cs2SnI6 for intermediate-band solar cells." *Materials Letters* 218 (2018): 233–236.

22. Owolabi, Taoreed O., Mohamed Faiz, Sunday O. Olatunji, and Idris K. Popoola. "Computational intelligence method of determining the energy band gap of doped ZnO semiconductor." *Materials & Design* 101 (2016): 277–284.

23. De Angelis, Filippo, and Lidia Armelao. "Optical properties of ZnO nanostructures: a hybrid DFT/TDDFT investigation." *Physical Chemistry Chemical Physics* 13, no. 2 (2011): 467–475.

24. Kurth, S., M. A. L. Marques, and E. K. U. Gross. "Density-Functional Theory." (2005): 395–402.

25. King, Burnham W., and Leon L. Suber. "Some properties of the oxides of vanadium and their compounds." *Journal of the American Ceramic Society* 38, no. 9 (1955): 306–311.

26. Morin, F. J. "Oxides which show a metal-to-insulator transition at the Neel temperature." *Physical Review Letters* 3, no. 1 (1959): 34.

27. Hyland, G. J. "On the electronic phase transitions in the lower oxides of vanadium." *Journal of Physics C: Solid State Physics* 1, no. 1 (1968): 189.

28. Lamsal, Chiranjivi, and N. M. Ravindra. "Optical properties of vanadium oxides—an analysis." *Journal of Materials Science* 48, no. 18 (2013): 6341–6351.

29. Nag, Joyeeta, Richard F. Haglund Jr, E. Andrew Payzant, and Karren L. More. "Non-congruence of thermally driven structural and electronic transitions in VO2." *Journal of Applied Physics* 112, no. 10 (2012): 103532.

30. Wu, B., A. Zimmers, H. Aubin, R. Ghosh, Y. Liu, and R. Lopez. "Electric-field-driven phase transition in vanadium dioxide." *Physical Review B* 84, no. 24 (2011): 241410.

31. Liu, Mengkun, Harold Y. Hwang, Hu Tao, Andrew C. Strikwerda, Kebin Fan, George R. Keiser, Aaron J. Sternbach et al. "Terahertz-field-induced insulator-to-metal transition in vanadium dioxide metamaterial." *Nature* 487, no. 7407 (2012): 345–348.

32. Aetukuri, Nagaphani B., Alexander X. Gray, Marc Drouard, Matteo Cossale, Li Gao, Alexander H. Reid, Roopali Kukreja et al. "Control of the metal–insulator transition in vanadium dioxide by modifying orbital occupancy." *Nature Physics* 9, no. 10 (2013): 661–666.

33. Yang, Zheng, Changhyun Ko, and Shriram Ramanathan. "Oxide electronics utilizing ultrafast metal-insulator transitions." *Annual Review of Materials Research* 41 (2011): 337–367.

34. Kakiuchida, Hiroshi, Ping Jin, Setsuo Nakao, and Masato Tazawa. "Optical properties of vanadium dioxide film during semiconductive–metallic phase transition." *Japanese Journal of Applied Physics* 46, no. 2L (2007): L113.

35. Xu, J-F., R. Czerw, S. Webster, D. L. Carroll, J. Ballato, and R. Nesper. "Nonlinear optical transmission in VOx nanotubes and VOx nanotube composites." *Applied Physics Letters* 81, no. 9 (2002): 1711–1713.

36. Wang, Weiping, Yongquan Luo, Dayong Zhang, and Fei Luo. "Dynamic optical limiting experiments on vanadium dioxide and vanadium pentoxide thin films irradiated by a laser beam." *Applied Optics* 45, no. 14 (2006): 3378–3381.

37. Soltani, M., M. Chaker, E. Haddad, R. V. Kruzelecky, and D. Nikanpour. "Optical switching of vanadium dioxide thin films deposited by reactive pulsed laser deposition." *Journal of Vacuum Science & Technology A: Vacuum, Surfaces, and Films* 22, no. 3 (2004): 859–864.

38. Jerominek, Hubert, Francis Picard, and Denis Vincent. "Vanadium oxide films for optical switching and detection." *Optical Engineering* 32, no. 9 (1993): 2092–2099.

39. Wan, Chenghao, Erik H. Horak, Jonathan King, Jad Salman, Zhen Zhang, You Zhou, Patrick Roney et al. "Limiting optical diodes enabled by the phase transition of vanadium dioxide." *ACS Photonics* 5, no. 7 (2018): 2688–2692.

40. Kats, Mikhail A., Romain Blanchard, Shuyan Zhang, Patrice Genevet, Changhyun Ko, Shriram Ramanathan, and Federico Capasso. "Vanadium dioxide as a natural disordered metamaterial: perfect thermal emission and large broadband negative differential thermal emittance." *Physical Review X* 3, no. 4 (2013): 041004.

41. Wu, Changzheng, Feng Feng, and Yi Xie. "Design of vanadium oxide structures with controllable electrical properties for energy applications." *Chemical Society Reviews* 42, no. 12 (2013): 5157–5183.

42. Liu, D., Y. Guo, L. Lin, and J. Robertson. "First-principles calculations of the electronic structure and defects of Al_2O_3." *Journal of Applied Physics* 114, no. 8 (2013): 083704.

43. Chain, Elizabeth E. "Optical properties of vanadium dioxide and vanadium pentoxide thin films." *Applied Optics* 30, no. 19 (1991): 2782–2787.

44. Imada, Masatoshi, Atsushi Fujimori, and Yoshinori Tokura. "Metal-insulator transitions." *Reviews of Modern Physics* 70, no. 4 (1998): 1039.

45. Jian, Jie, Aiping Chen, Youxing Chen, Xinghang Zhang, and Haiyan Wang. "Roles of strain and domain boundaries on the phase transition stability of VO_2 thin films." *Applied Physics Letters* 111, no. 15 (2017): 153102.

46. Zhao, Yong, Joon Hwan Lee, Yanhan Zhu, M. Nazari, Changhong Chen, Haiyan Wang, Ayrton Bernussi, Mark Holtz, and Zhaoyang Fan. "Structural, electrical, and terahertz transmission properties of VO_2 thin films grown on c-, r-, and m-plane sapphire substrates." *Journal of Applied Physics* 111, no. 5 (2012): 053533.

47. Rensberg, Jura, Shuyan Zhang, You Zhou, Alexander S. McLeod, Christian Schwarz, Michael Goldflam, Mengkun Liu et al. "Active optical metasurfaces based on defect-engineered phase-transition materials." *Nano Letters* 16, no. 2 (2016): 1050–1055.

48. Wan, Chenghao, Zhen Zhang, David Woolf, Colin M. Hessel, Jura Rensberg, Joel M. Hensley, Yuzhe Xiao et al. "On the optical properties of thin-film vanadium dioxide from the visible to the far infrared." *Annalen der Physik* 531, no. 10 (2019): 1900188.

49. Parker, J. C., D. J. Lam, Y-N. Xu, and W. Y. Ching. "Optical properties of vanadium pentoxide determined from ellipsometry and band-structure calculations." *Physical Review B* 42, no. 8 (1990): 5289.

50. Held, Karsten, Georg Keller, Volker Eyert, Dieter Vollhardt, and Vladimir I. Anisimov. "Mott-Hubbard metal-insulator transition in paramagnetic V_2O_3: an LDA + DMFT (QMC) study." *Physical Review Letters* 86, no. 23 (2001): 5345.

51. Baldassarre, L., A. Perucchi, D. Nicoletti, A. Toschi, G. Sangiovanni, K. Held, Massimo Capone et al. "Quasiparticle evolution and pseudogap formation in V_2O_3: an infrared spectroscopy study." *Physical Review B* 77, no. 11 (2008): 113107.

52. Tomczak, Jan M., and Silke Biermann. "Multi-orbital effects in optical properties of vanadium sesquioxide." *Journal of Physics: Condensed Matter* 21, no. 6 (2009): 064209.

53. McWhan, D. B., T. M. Rice, and J. P. Remeika. "Mott transition in Cr-doped V_2O_3." *Physical Review Letters* 23, no. 24 (1969): 1384.

54. Biermann, Silke, A. Poteryaev, A. I. Lichtenstein, and A. Georges. "Dynamical singlets and correlation-assisted Peierls transition in VO_2." *Physical Review Letters* 94, no. 2 (2005): 026404.

55. Tomczak, Jan M., and Silke Biermann. "Materials design using correlated oxides: optical properties of vanadium dioxide." *EPL (Europhysics Letters)* 86, no. 3 (2009): 37004.

56. Marini, C., E. Arcangeletti, D. Di Castro, L. Baldassare, A. Perucchi, S. Lupi, L. Malavasi et al. "Optical properties of $V_{1-x}Cr_xO_2$ compounds under high pressure." *Physical Review B* 77, no. 23 (2008): 235111.

57. Schilbe, Peter. "Raman scattering in VO_2." *Physica B: Condensed Matter* 316 (2002): 600–602.

58. El Haimeur, A., A. Mrigal, H. Bakkali, L. El Gana, K. Nouneh, M. Addou, and M. Dominguez. "Optical, magnetic, and electronic properties of nanostructured VO_2 thin films grown by spray pyrolysis: DFT first principle study." *Journal of Superconductivity and Novel Magnetism* 33, no. 2 (2020): 511–517.

59. Zayed, M. K., A. A. Elabbar, and O. A. Yassin. "Electronic and optical properties of the VO_2 monoclinic phase using SCAN meta-GGA and TB-mBJ methods." *Physica B: Condensed Matter* 582 (2020): 411887.

60. Szymanski, N. J., Z. T. Y. Liu, T. Alderson, N. J. Podraza, P. Sarin, and S. V. Khare. "Electronic and optical properties of vanadium oxides from first principles." *Computational Materials Science* 146 (2018): 310–318.

61. Vecchio, I. Lo, M. Autore, F. D'Apuzzo, F. Giorgianni, A. Perucchi, U. Schade, V. N. Andreev, V. A. Klimov, and S. Lupi. "Optical conductivity of V 4 O 7 across its metal-insulator transition." *Physical Review B* 90, no. 11 (2014): 115149.

62. Qazilbash, M. M.; Schafgans, A.; Burch, K.; Yun, S.; Chae, B.; Kim, B.; Kim, H.-T. & D., Basov. "Electrodynamics of the vanadium oxides VO_2 and V_2O_3." *Physical Review B, APS*, 77 (2008): 115121.

63. Kang, Tae Dong, J-S. Chung, and Jong-Gul Yoon. "Anisotropic optical response of nanocrystalline V_2O_5 thin films and effects of oxygen vacancy formation." *Physical Review B* 89, no. 9 (2014): 094201.

64. Baldassarre, L., A. Perucchi, E. Arcangeletti, D. Nicoletti, D. Di Castro, P. Postorino, V. A. Sidorov, and S. Lupi. "Electrodynamics near the metal-to-insulator transition in V_3O_5." *Physical Review B* 75, no. 24 (2007): 245108.

65. Irizawa, Akinori, A. Higashiya, M. Tsunekawa, A. Sekiyama, S. Imada, S. Suga, T. Yamauchi et al. "Metal–insulator transition in V6O13 probed by photoemission and optical studies." *Journal of Electron Spectroscopy and Related Phenomena* 144 (2005): 345–347.

66. Ambrosini, Andrea, Angel Duarte, Kenneth R. Poeppelmeier, Melissa Lane, Carl R. Kannewurf, and Thomas O. Mason. "Electrical, optical, and structural properties of tin-doped In_2O_3–M_2O_3 solid solutions (M= Y, Sc)." *Journal of Solid State Chemistry* 153, no. 1 (2000): 41–47.
67. Hamberg, I., A. Hjortsberg, and C. G. Granqvist. "High quality transparent heat reflectors of reactively evaporated indium tin oxide." *Applied Physics Letters* 40, no. 5 (1982): 362–364.
68. Hamberg, Ivar, Claes Göran Granqvist, K-F. Berggren, Bo E. Sernelius, and L. Engström. "Band-gap widening in heavily Sn-doped In_2O_3." *Physical Review B* 30, no. 6 (1984): 3240.
69. Brewer, Scott H., and Stefan Franzen. "Indium tin oxide plasma frequency dependence on sheet resistance and surface adlayers determined by reflectance FTIR spectroscopy." *The Journal of Physical Chemistry B* 106, no. 50 (2002): 12986–12992.
70. Hjortsberg, A., I. Hamberg, and C. G. Granqvist. "Transparent and heat-reflecting indium tin oxide films prepared by reactive electron beam evaporation." *Thin Solid Films* 90, no. 3 (1982): 323–326.
71. Kaplan, L., A. Ben-Shalom, R. L. Boxman, S. Goldsmith, U. Rosenberg, and M. Nathan. "Annealing and Sb-doping of Sn—O films produced by filtered vacuum arc deposition: structure and electro-optical properties." *Thin Solid Films* 253, no. 1-2 (1994): 1–8.
72. Yan, Yanfa, J. Zhou, X. Z. Wu, H. R. Moutinho, and M. M. Al-Jassim. "Structural instability of Sn-doped In_2O_3 thin films during thermal annealing at low temperature." *Thin Solid Films* 515, no. 17 (2007): 6686–6690.
73. Lai, Fachun, Limei Lin, Rongquan Gai, Yongzhong Lin, and Zhigao Huang. "Determination of optical constants and thicknesses of In_2O_3:Sn films from transmittance data." *Thin Solid Films* 515, no. 18 (2007): 7387–7392.
74. Hong, Hyun-Gi, Seok-Soon Kim, Dong-Yu Kim, Takhee Lee, June-O. Song, J. H. Cho, C. Sone, Y. Park, and Tae-Yeon Seong. "Enhancement of the light output of GaN-based ultraviolet light-emitting diodes by a one-dimensional nanopatterning process." *Applied Physics Letters* 88, no. 10 (2006): 103505.
75. Bregman, J., Yoram Shapira, and H. Aharoni. "Effects of oxygen partial pressure during deposition on the properties of ion-beam-sputtered indium-tin oxide thin films." *Journal of Applied Physics* 67, no. 8 (1990): 3750–3753.
76. Fan, John C. C., Frank J. Bachner, and George H. Foley. "Effect of O_2 pressure during deposition on properties of rf-sputtered Sn-doped In_2O_3 films." *Applied Physics Letters* 31, no. 11 (1977): 773–775.
77. Shin, J. H., S. H. Shin, J. I. Park, and H. H. Kim. "Properties of dc magnetron sputtered indium tin oxide films on polymeric substrates at room temperature." *Journal of Applied Physics* 89, no. 9 (2001): 5199–5203.
78. Qiao, Z., R. Latz, and D. Mergel. "Thickness dependence of In2O3: Sn film growth." *Thin Solid Films* 466, no. 1-2 (2004): 250–258.
79. Brewer, Scott H., and Stefan Franzen. "Optical properties of indium tin oxide and fluorine-doped tin oxide surfaces: correlation of reflectivity, skin depth, and plasmon frequency with conductivity." *Journal of Alloys and Compounds* 338, no. 1-2 (2002): 73–79.
80. Brewer, Scott H., and Stefan Franzen. "Calculation of the electronic and optical properties of indium tin oxide by density functional theory." *Chemical Physics* 300, no. 1-3 (2004): 285–293.
81. Segall, M. D., Philip J. D. Lindan, M. J. al Probert, Christopher James Pickard, Philip James Hasnip, S. J. Clark, and M. C. Payne. "First-principles simulation: ideas, illustrations and the CASTEP code." *Journal of Physics: Condensed Matter* 14, no. 11 (2002): 2717.
82. Perdew, John P., Kieron Burke, and Matthias Ernzerhof. "Generalized gradient approximation made simple." *Physical Review Letters* 77, no. 18 (1996): 3865.
83. Troullier, Norman, and José Luís Martins. "Efficient pseudo-potentials for plane-wave calculations." *Physical Review B* 43, no. 3 (1991): 1993.
84. Qin, Guoqiang, Dongchun Li, Zhijun Feng, and Shimin Liu. "First principles study on the properties of p-type conducting In:SnO_2." *Thin Solid Films* 517, no. 11 (2009): 3345–3349.
85. Tripathi, Madhvendra Nath, Kazuhito Shida, Ryoji Sahara, Hiroshi Mizuseki, and Yoshiyuki Kawazoe. "First-principles analysis of structural and opto-electronic properties of indium tin oxide." *Journal of Applied Physics* 111, no. 10 (2012): 103110.
86. Bai, L. N., Y. P. Wei, J. S. Lian, and Q. Jiang. "Stability of indium–tin-oxide and its optical properties: a first-principles study." *Journal of Physics and Chemistry of Solids* 74, no. 3 (2013): 446–451.
87. Liu, Xiaoyan, Lei Wang, and Yi Tong. "Optoelectronic properties of ultrathin indium oxide films: a first-principle study." *Crystals* 11, no. 1 (2021): 30.

Index

A

AAO templates, *see* Anodized aluminium oxide templates
Absorption coefficient, 108
Acoustic cavitation, 15
Advanced oxidation technologies (AOT), 13
Agriculture, semiconductor optical utilization in, 39
    semiconductor photocatalyst, 40–41
    silica nanoparticles (SiNPs), 43–44
    silver nanoparticles, 42–43
    titanium dioxide (TiO_2) nanomaterials, 41–42
    zinc oxide nanoparticles, 44–46
Amorphous silicon, 96
Anodized aluminium oxide (AAO) templates, 77
AOT, *see* Advanced oxidation technologies
Arc discharge, 15
ATR, *see* Attenuated total reflection
Attenuated total reflection (ATR), 144
Attenuation, 8
Audio compressors, 81–82
Automatic street lighting, 83–84

B

Ball milling, 19
Bandgap, 21, 39, 40
Band structure theory, assumptions and limits of, 136
Beer-Lambert Law, 81
Beer's law, 66
Benzil (Bz), 69
Bethe-Salpeter equation (BSE), 155
Bi_2O_3-based photocatalysts, 112
Biexcitons, 53
Binary III–V semiconductor core fiber, 6
Bio-recalcitrant organic compounds (BROC), 13
Bipolaris sorokiniana, 42
Bismuth oxide, 112
BLA, *see* Bond length alternation
Bond length alternation (BLA), 70
Bouguer-Lambert's law, 108, 110
Brillouin zone, 138
BROC, *see* Bio-recalcitrant organic compounds
BSE, *see* Bethe-Salpeter equation
Bulk magnetic semiconductors, magnetic field on, 150
Bulk organic semiconductors, 69
Bulk semiconductors, application of electric field on, 148

C

Cadmium telluride and cadmium sulfide, 97
Cambridge Serial Total Energy Package (CASTEP)
    code, 160
Carbon-based compounds, 68
Carbon nanotube (CNT), 15, 76
Carbon nitride, 113
CASTEP code, *see* Cambridge Serial Total Energy
    Package code
CB, *see* Conduction band

CdS-based photocatalysts, 113
CG approximation, *see* Conjugate gradient approximation
Chemical vapor deposition (CVD), 19
Chromium, 117
Clemson University, 2, 6
Close spaced sublimation (CSS) technique, 97
CMOS, *see* Complementary metal-oxide semiconductors
CNT, *see* Carbon nanotube
Coherent potential approximation (CPA), 160
Complementary metal-oxide semiconductors (CMOS), 84
Conduction band (CB), 14, 15, 136
Conjugate gradient (CG) approximation, 31
Coulomb screening, 53
CPA, *see* Coherent potential approximation
Crystalline silicon, 96–97
Crystalline silicon on glass (CSG), 96
CSG, *see* Crystalline silicon on glass
CSS technique, *see* Close spaced sublimation technique
CuS-based photocatalysts, 113
Cutoff wavelength, 8
CVD, *see* Chemical vapor deposition

D

DANS, *see* 4-(N, N dimethyl amino)-4´-nitrostilbene
Dember effect, 110
Density functional theory (DFT), 31, 157–158
Density of states (DOS), 24
DFG, *see* Difference frequency generation
DFT, *see* Density functional theory
Dielectric constant, 35
Difference frequency generation (DFG), 52
Diluted semiconductor (DMS), 150
4-(N,N Dimethyl amino)-4´-nitrostilbene (DANS), 70
Dispersion, 8
DMFT, *see* Dynamical mean field approximation
DMS, *see* Diluted semiconductor
DOS, *see* Density of states
Double porous silicon (d-PS) layers deposition, 96
d-PS layers deposition, *see* Double porous silicon layers
    deposition
DSSC technology, *see* Dye-sensitized solar cell technology
Dye photosensitization on surface, 119–120
Dye-sensitized solar cell (DSSC) technology, 99
Dynamical mean field approximation (DMFT), 159

E

Effective-medium-theory (EMT), 159
EHP, *see* Electron-hole plasma
Electric field effect on excitons and carriers, 148
    on bulk semiconductors, 148
    on low-dimensional structure, 148–149
Electrodeposition, 20
Electromagnetic (EM) radiations, 92
    in matter, 131–134
    in vacuum, 130–131
Electron-hole plasma (EHP), 53

For Product Safety Concerns and Information please contact our EU
representative GPSR@taylorandfrancis.com
Taylor & Francis Verlag GmbH, Kaufingerstraße 24, 80331 München, Germany

www.ingramcontent.com/pod-product-compliance
Lightning Source LLC
Chambersburg PA
CBHW081530220326
41598CB00036B/6382